CHEMISTRY RESEARCH AND APPLICATIONS

ALCOHOL OXIDATION

REACTION, EFFECTS AND APPLICATIONS

Chemistry Research and Applications

Additional books and e-books in this series can be found on Nova's website under the Series tab.

CHEMISTRY RESEARCH AND APPLICATIONS

ALCOHOL OXIDATION

REACTION, EFFECTS AND APPLICATIONS

STEFFAN PATERSON
EDITOR

nova
science publishers
New York

Library of Congress Cataloging-in-Publication Data

ISBN: 978-1-53614-604-2

Published by Nova Science Publishers, Inc. † New York

CONTENTS

PREFACE

Over the last few years, there has been a growing interest in the development of sustainable processes for large-scale production of commodities. As such, Alcohol Oxidation: Reaction, Effects and Applications addresses recent developments in metal-catalyzed selective oxidation reactions of alcohols to carbonyl-containing compounds under unconventional conditions.

Following this, the authors examine how simple alcohols have been used as feedstocks for the preparation of a variety of enantio-enriched substances. Cascade reactions allow for the elaboration of complex molecules from simple starting materials in a very efficient way. Several bonds can be made in one-pot procedures without the need for intermediates to be isolated.

Metal-organic frameworks and covalent organic frameworks are two classes of highly porous materials that have been heavily investigated in recent years. As such, in this compilation the authors focus on the use of metal-organic frameworks as heterogeneous oxidation catalysts.

Enantiopure alcohols are important starting materials and intermediates for the synthesis of many biologically active compounds and fine chemicals. The use of enzymes for the kinetic resolution of racemates has produced affordable enantiopure compounds in good yields, however non-enzymatic chiral catalytic systems have been gaining ground. Accordingly, the effective non-enzymatic oxidative kinetic resolutions of racemic alcohols and related compounds such as diols or benzoins are summarized and discussed.

Lastly, the potential employment of the electrochemical process to oxidize alcohols in a more efficient way than the usual chemical method is presented. The use of the electrode potential as a controllable variable and driving force of the process allows for the oxidation of some alcohol molecules without the very aggressive nature of reactants.

Chapter 1 - Over the last few years, there has been a growing interest on the development of sustainable processes for large-scale production of commodities. This chapter addresses recently achieved sustainable developments on metal-catalyzed selective oxidation reactions of alcohols to carbonyl-containing compounds under unconventional conditions. The proposed mechanisms for the catalytic oxidative processes are also outlined.

Chapter 2 - Cascade reactions allow the elaboration of complex molecules from simple starting materials in a very efficient way. Several bonds can be made in one-pot procedures without the need for intermediates to be isolated. In this chapter the authors examine how simple alcohols have been used in this manner as feedstocks for the preparation of a variety of enantioenriched substances.

Chapter 3 - Metal-Organic Frameworks (MOF, invented in 1999) and Covalent Organic Frameworks (COF, invented in 2005) are two classes of highly porous and designable materials that have been heavily investigated over the last years.

A MOF typically consists of inorganic nodes that are connected by organic linkers to form crystalline, highly porous structures. In this chapter, the authors will focus on the use of MOFs as heterogeneous oxidation catalysts. MOFs are very promising candidates to replace homogeneous catalysts with sustainable and stable heterogeneous catalysts. The catalytic active function can be either the active metal sites of the MOF itself or can be introduced as an extra functionality in the linker, a dopant or a ship-in-a-bottle complex. As the pore size, pore shape and functionality of MOFs can be designed in numerous ways, shape selectivity and even chiral selectivity can be created.

COFs can be referred to as metal-free porous polymers, except the fact that COFs are (normally) crystalline materials. Many structures have been developed in the last few years, from the original boronate COFs to imine COFs, and CTFs (Covalent Triazine Frameworks) and many others. In the field of oxidation catalysis, the stability is obviously very important. Imine COFs and CTFs are much more stable than MOFs, but their tunability is somewhat lower and the synthesis of many COFs remains a bit cumbersome. However, they open a path to heterogeneous metal-free (photo)catalysis, and that is a very promising route.

Chapter 4 - Enantiopure alcohols are important starting materials and intermediates for the synthesis of many biologically active compounds and fine chemicals. One of the simplest and most convenient ways to access them is an oxidative kinetic resolution of racemic alcohols. The use of enzymes for the kinetic resolution of racemates has given affordable enantiopure compounds in good yields, however, non-enzymatic chiral catalytic systems have been gaining ground. Accordingly, the effective non-enzymatic oxidative kinetic resolutions of racemic alcohols and related compounds such as diols or benzoins (α-hydroxy ketones) are summarized and discussed in this chapter.

Chapter 5 - In this chapter, the potential employment of the electrochemical process to oxidize alcohols in a more efficient way than the usual chemical method is presented. The use of the electrode potential as a controllable variable and driving force of the process allows for the oxidation of some alcohol molecules without of the very aggressive nature of reactants. The use of alcohols as fuel in fuel cells opens a wide field of improvements for energy and added value chemicals production. Methanol and ethanol electrooxidation are discussed and very promising results obtained from ethylene glycol and glycerol reactions show the possible use of these alcohols as precursors for energy production in a sustainable development concept as well as the production of chemicals of interest for industries.

In: Alcohol Oxidation
Editor: Steffan Paterson

ISBN: 978-1-53614-604-2

Chapter 1

ALCOHOL OXIDATION UNDER UNCONVENTIONAL REACTION CONDITIONS

Luísa M. D. R. S. Martins*

Centro de Química Estrutural, Instituto Superior Técnico,
Universidade de Lisboa, Lisboa, Portugal

ABSTRACT

Over the last few years, there has been a growing interest on the development of sustainable processes for large-scale production of commodities. This chapter addresses recently achieved sustainable developments on metal-catalyzed selective oxidation reactions of alcohols to carbonyl-containing compounds under unconventional conditions. The proposed mechanisms for the catalytic oxidative processes are also outlined.

Keywords: oxidation, alcohol, catalyst, microwave, ionic liquid, metal-complex, ultrasound

---

* Corresponding Author Email:luisammartins@tecnico.ulisboa.pt.

INTRODUCTION

The oxidation of alcohols to carbonyl-containing compounds are among the central reactions in industrial organic chemistry and involved in the development of environmentally benign processes and production of new materials [1-5]. Advances on new procedures, catalysts, oxidation agents, and uses for alcohols oxidation have been regularly surveyed [3, 6] being one of the most extensively investigated areas of current organic synthesis. In particular, the partial oxidation of benzyl alcohol and of methylbenzyl alcohol (low toxic naturally produced aromatic alcohols), selectively originating benzaldehyde and acetophenone, respectively (Scheme 1), has a great importance in industrial organic synthesis as precursors to other organic compounds, ranging from pharmaceuticals to plastic additives.

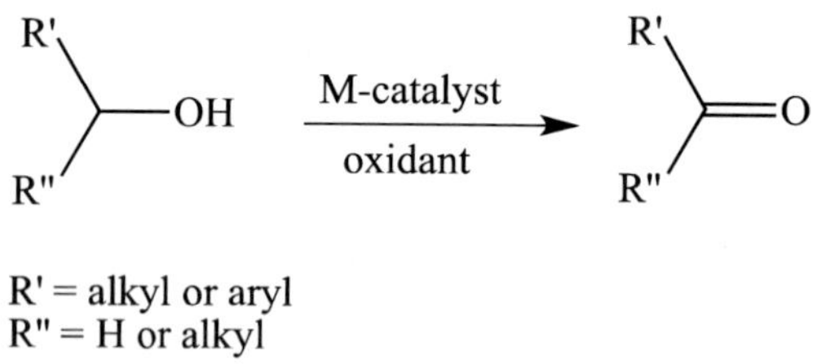

Scheme 1. Metal-catalyzed oxidation of primary or secondary alcohols.

However, alcohol oxidation still constitutes a challenging field of research to develop more effective catalysts or search for more advantageous oxidation conditions [1-5]. Thus, currently, one of the aims of metal complex catalysis is the search for new efficient systems able to selectively oxidize alcohols under sustainable conditions.

Many homogeneous catalytic systems, in spite of being very active and selective, frequently cannot lead to a practical solution of technological problems due to their high cost, instability or difficulty of recovery and recycling. Hence, efforts are been devoted to obtaining recyclable catalytic systems either by supporting active homogeneous catalysts or designing new structures (such as metal organic frameworks or coordination polymers), which, even if being less active than the homogeneous ones in a single use, can be re-used several times [3, 6, 7].

Application of low energy irradiation such as microwaves [8-11] or ultrasounds [12, 13] is also among the new alternative techniques which can shorten the reaction time in oxidations of alcohols. Moreover, ongoing approaches on the development of new environmentally friendly protocols for alcohols oxidation involves the combination of microwave and ultrasonic irradiation in room temperature ionic liquids [14]. The use of supercritical conditions, in particular the low toxic, stable supercritical CO_2 fluid, to enhance alcohols oxidation processes is also emerging [15].

In the following sections a general overview of energetically sustainable one-pot metal-catalyzed oxidation of primary or secondary alcohols (Scheme 1) under unconventional media (e.g., under microwave irradiation or in ionic liquids) is unfolded.

METHODS

It is widely recognized that the use of low energy microwave (MW) irradiation has great advantages on promoting and accelerating chemical reactions by reducing reaction time, allowing solvent-free conditions and increasing yields of valued-added compounds [8-11]. MW-assisted reactions are based on the efficient interaction of reactants with electromagnetic waves generated by a microwave dielectric effect which depends on the ability of such reactants to absorb microwave energy and convert it into heat. Polar substances, having good potential to perform such conversion, accelerate MW-induced reactions (in comparison when submitted to conventional heating). In fact, despite alcohols have relative low dielectric constant they heat rapidly in a MW field due to their high loss tangent (tan δ, the ability of a specific material to convert electromagnetic energy into heat) required for efficient absorption and consequently rapid heating.

Moreover, MW-enhanced catalytic reactions can be performed under moderate conditions with high selectivity for the desired products (selectivity, ratio of the number of moles of the desired product (aldehyde or ketone) formed by the number of moles of reacted substrate) [16] in a sustainable environment.

The accelerating effect of MW irradiation in the synthesis of ketones from secondary alcohols (Scheme 1, R" = alkyl) with *tert*-butyl hydroperoxide (*t*-BuOOH, TBHP) as oxidant is one of the most reported examples of successful application of microwaves in catalytic oxidation reactions of alcohols [3, 17-27]. The preferred model substrate is methylbenzyl alcohol (also designed as 1-phenylethanol), leading to the formation of acetophenone, in most cases, in quantitative yields. A wide range of metal complexes can act as catalysts or catalyst precursors for this MW-assisted model reaction. Thus, the given examples will be classified according to the metal center of the complex.

Oxidovanadium species in which vanadium is in the +5 oxidation state (e.g., VO^{3+}, VO_2^+, $V_2O_3^{4+}$, and $V_2O_4^{2+}$) exhibit a hard acidic nature yielding to a variety of oxidovanadium(V) complexes with *N,O*-donor ligands, able to effectively catalyze the MW-enhanced oxidation of primary or secondary alcohols (Scheme 1) [3, 17, 20, 21], either in liquid or supported (solid) systems.

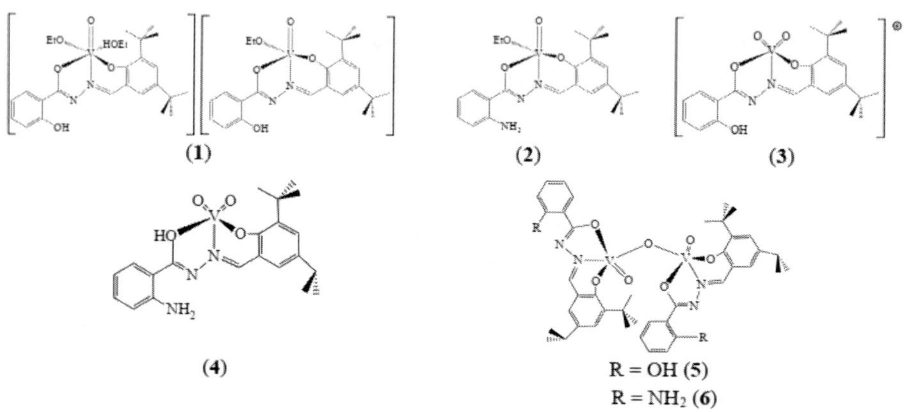

Figure 1. Oxidovanadium(V) complexes [VOL1(OEt)]·[VOL1(OEt)(EtOH)] (1), [VOL2(OEt)] (2), [Et3NH][VO2L1] (3), [VO2(HL2)]·2EtOH (4), [(VOL1)2(μ-O)] (5) and [(VOL2)2(μ-O)] (6).

For example, the catalytic activity of oxidovanadium(V) aroylhydrazone complexes derived from (3,5-di-*tert*-butyl-2-hydroxybenzylidene)-2-hydroxybenzohydrazide (H_2L^1) and (3,5-di-*tert*-butyl-2-hydroxybenzylidene)-2-aminobenzohydrazide (H_2L^2), *viz.* [VOL^1(OEt)]·[VOL^1(OEt)(EtOH)] (**1**), [VOL^2(OEt)] (**2**), [Et_3NH][VO_2L^1] (**3**), [$VO_2(HL^2)$]·2EtOH (**4**), [$(VOL^1)_2$(μ-O)] (**5**) and [$(VOL^2)_2$(μ-O)] (**6**) (Figure 1) was tested for the oxidation of the secondary alcohol 1-phenylethanol to acetophenone, in mild green reaction conditions [20].

The authors found that **1–6** are active catalysts in the microwave-assisted solvent-free peroxidative (by *t*-BuOOH) oxidation of 1-phenylethanol both in homogeneous and heterogeneous conditions. Activated carbon (AC), multi-walled carbon nanotubes (MWCNT), nanodiamonds (ND) and carbon xerogel (CX) were used as supports in their original forms, as purchased (AC, ND and CNT) or as prepared (CX), oxidized (-ox), and oxidized with nitric acid and subsequently treated with sodium hydroxide (-ox-Na). The different support treatments affected the structure and surface chemistry of the carbon materials, but all were able to anchor the oxidovanadium complexes **1** - **6**, although with different efficiencies [20]. The authors believe that for complexes **1** and **2** (used as models, Figure 2) the coordination of the complex to the functionalized carbon material could occur *i*) at the vacant 6th position through the –OH or –COO^- group at the surface of the CNTs, *ii*) by replacing coordinated ethoxido group by –OH group on the carbon materials, *iii*) via hydrogen bonding or *iv*) via ionic interactions. In case of ionic complexes (**3** and **4**) hydrogen bonding interactions should be more effective. For the μ-oxido-bridged oxidovanadium(V) complexes **5** and **6** all above modes are possible except the replacement of ethoxido group (as they don't have it) [20].

In homogeneous conditions, almost quantitative conversion, up to 94% (**5**), of 1-phenylethanol in acetophenone (>98% selectivity) was obtained with a turnover frequency (TOF, moles of product per mol of catalyst) per hour] of 9.4×10^2 h^{-1}, after 1 h of MW irradiation at 90 °C in the presence of 0.1 mol% (*vs.* substrate) of the oxidovanadium (V) compound [20].

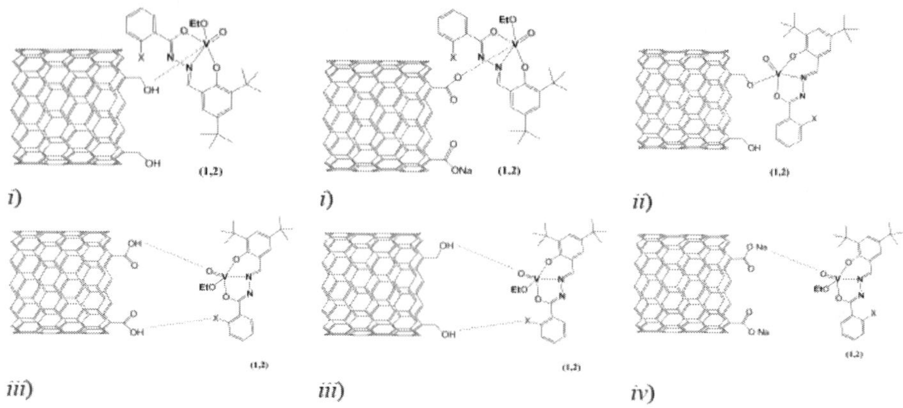

Figure 2. Possible ways of anchoring oxidovanadium (V) complexes on functionalized multi-walled carbon nanotubes.

The above peroxidative oxidation of 1-phenylethanol appears to proceed *via* a radical mechanism (equations 1 – 6) as previously proposed for other systems [17, 19, 21]. It may involve the vanadium-assisted generation of $^tBuOO^{\cdot}$ and $^tBuO^{\cdot}$ radicals according to equations (1) and (1A) or (1B) and (2), respectively, which behave as hydrogen atom abstractors from the alcohol [equations (4) and (5)].

$$VO^{3+} + {}^tBuOOH \rightarrow VO^{2+} + {}^tBuOO^{\cdot} + H^+ \quad (1)$$

$$VO_2^{+} + {}^tBuOOH \rightarrow VO^{2+} + {}^tBuOO^{\cdot} + OH^- \quad (1A)$$

$$V_2O_3^{4+} + {}^tBuOOH \rightarrow VO^{2+} + VO^{3+} + {}^tBuOO^{\cdot} + OH^- \quad (1B)$$

$$VO^{2+} + {}^tBuOOH \rightarrow VO^{3+}\text{–}(OH^-) + {}^tBuO^{\cdot} \quad (2)$$

$$VO^{3+}\text{–}(OH^-) + {}^tBuOOH \rightarrow VO^{3+}\text{–}(OO\text{–}{}^tBu^-) + H_2O \quad (3)$$

$${}^tBuO^{\cdot} + PhCH(OH)CH_3 \rightarrow {}^tBuOH + PhC^{\cdot}\text{–}(OH)CH_3 \quad (4)$$

$${}^tBuOO^{\cdot} + PhCH(OH)CH_3 \rightarrow {}^tBuOOH + PhC^{\cdot}\text{–}(OH)CH_3 \quad (5)$$

$$VO^{3+}\text{–}(OO\text{–}{}^tBu^-) + PhC^{\cdot}\text{–}OHCH_3 \rightarrow Ph(C{=}O)CH_3 + {}^tBuOOH + VO^{2+} \quad (6)$$

The immobilization of the oxidovanadium complexes **1**, **2**, 4 and **5** on functionalized carbon materials (e.g., carbon xerogels, activated carbons, carbon nanotubes or nanodiamonds) improved the MW-induced oxidation efficiency and allowed catalyst reuse while preserving their activity and selectivity [20].

Examples of heterogenization of Fe-catalysts to be used as heterogeneous catalysts for the peroxidative (with TBHP) oxidation of alcohols while taking advantages of the features of homogeneous catalytic systems, include the bio-inspired tris(pyrazol-1-yl)methane C-scorpionate Fe(II) complex $[FeCl_2\{\kappa^3\text{-}HC(pz)_3\}]$ (**7**, pz = pyrazol-1-yl) supported on functionalized multi-walled carbon nanotubes (**7**@MWCNT) as catalyst in a single-pot, solvent-free MW-promoted protocol [22]. The immobilization of **7** on the MWCNTs is believed to occur by interaction of the iron(II) center with the oxygenated functional groups of the MWCNTs surface [23], with formation of a covalent Fe-O bond, providing additional stability of the material in the reaction medium. This was the first time that a C-scorpionate complex was reported as catalyst for alcohol oxidation, either in homogenous or heterogeneous conditions [22].

Quantitative yields of both cyclohexanone and acetophenone were achieved after 1 h of MW-assisted oxidation of the corresponding alcohols with TBHP, in the presence of the **7**@MWCNT hybrid material and of 2,2,6,6-tetramethylpiperidine-*N*-oxyl radical (TEMPO, a widely used promoter for the aerobic oxidation of alcohols to the corresponding carbonyl compounds [1-7, 28, 29] but with few applications for the peroxidative oxidation of alcohols) [30-33].

7@MWCNT maintained its catalytic efficiency to produce cyclohexanone (with a yield over 98%) during five consecutive cycles and still retained 80% of its initial activity and concomitant selectivity at the 12th cycle. Thus, the immobilization of **7** at functionalized MWCNTs not only provided a more stable, easy to separate and reusable catalyst, but also remarkably improved the catalytic performance of **7** in the homogeneous oxidation of secondary alcohols (only 49% or 70% yields of cyclohexanone or acetophenone from cyclohexanol or 1-phenylethanol, respectively, were achieved after 1 h of MW irradiation in the homogeneous system) [22].

The scope of carbon supports for the C-scorpionate Fe(II) complex $[FeCl_2\{\kappa^3\text{-}HC(pz)_3\}]$ (**7**) has been very recently extended [24] to five different nanostructured carbon materials (Figure 3): nanodiamonds (ND, used for the first time as supports for iron complexes), graphene nanoplatelets (G-NPL), graphene oxide (GO), reduced graphene oxide (rGO) and single-walled nanohorns (SWCNH, the first ever reported complex supported on carbon nanohorns).

Under the optimized conditions [24], 1 h of MW irradiation (25 W) at 80 °C, 97.1% yield of acetophenone and 96.9% yield of 2-phenylacetaldehyde, respectively, were obtained from the selective MW-induced oxidation of 1- and 2-phenylethanol in the presence of 0.1 mol% vs. substrate of **7** supported at nanodiamonds (ND) pretreated with nitric acid (-ox) and subsequently with base (-Na) with TOFs of $9.7 \times 10^2\ h^{-1}$. The best results were obtained with the oxidized materials, confirming the importance of surface oxygen groups as anchorage sites. Moreover, **7**@ND-ox-Na was able to be reused for seven consecutive cycles without losing its catalytic activity (after the sixth cycle, 90% of the initial activity of **7**@ND-ox-Na was retained).

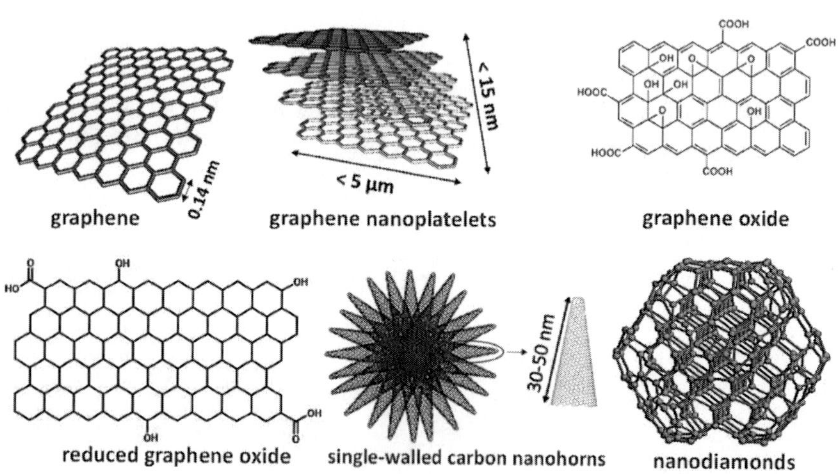

Figure 3. Carbon materials used as supports for [FeCl2{κ3-HC(pz)3}] (7).

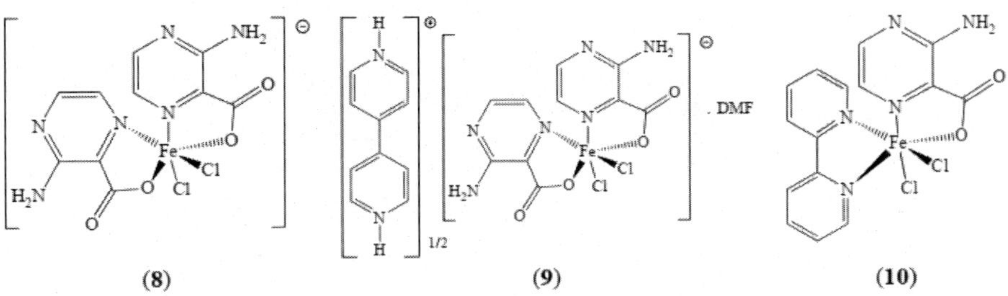

Figure 4. Iron(III) complexes [H(EtOH)][FeCl2(L)2] (8), [H2bipy]1/2[FeCl2(L)2].DMF (9) and [FeCl2(L)(2,2'-bipy)] (10).

Iron(III) complexes, namely [H(EtOH)][$FeCl_2(L)_2$] (**8**), [H_2bipy]$_{1/2}$ [$FeCl_2(L)_2$].DMF (**9**) and [$FeCl_2$(L)(2,2'-bipy)] (**10**) (L = 3-amino-2-pyrazinecarboxylate; H_2bipy = doubly protonated 4,4'-bipyridine; 2,2'-bipy = 2,2'-bipyridine, DMF = dimethylformamide) [25] (Figure 4) also exhibited high catalytic activity for the low power solvent-free microwave-assisted peroxidative oxidation of methyl benzyl alcohol, leading, in the presence of TEMPO, to quantitative yields of acetophenone [TOFs up to 8.1 $\times 10^3$ h^{-1} (**10**)] after 1 h of irradiation. Moreover, the catalysts are easy to recover (by filtration) and were reused, at least for four consecutive cycles, maintaining 83% of the initial activity and concomitant high selectivity [25].

Complexes **8–10** were also tested towards the oxidation of C_6 secondary alcohols, namely cyclohexanol and the linear 2- and 3-hexanol, and the primary alcohols benzyl alcohol and 1-hexanol. As expected [26, 27], the alicyclic cyclohexanol is less reactive than 1-phenylethanol: in the presence of a catalytic amount of **8–10** (0.4 mol% vs. substrate) and in the solvent-free medium, the yields of cyclohexanone are in the 61-66% range (after 1.5 h at 150 °C / 10W MW irradiation). The linear aliphatic alcohols, 2-hexanol and 3-hexanol, lead to lower yields in the 45-56% range, under similar reaction conditions, as previously reported [3, 17]. The activity of **8–10** in the oxidation of benzyl alcohol or 1-hexanol is quite modest since only moderated yields of benzaldehyde (30-34% range) or hexanal (22-24% range) are obtained under the same reaction conditions. Thus, the exhibited activity follows the order: methyl benzyl alcohol > cyclohexanol > 2-hexanol > 3-hexanol > benzyl alcohol > 1-hexanol.

Ruthenium- and rhodium-catalyzed hydrogen-transfer type oxidation of secondary alcohols (e.g., 4-*t*-butylcyclohexanol, cyclooctanol, cyclododecanol, 1-*p*-tolyl-1-hexanol or 5-tetradecanol,) is known [34]. Moderate to excellent yields, to the corresponding ketones were achieved by employing MW heating at 140 °C for 15 min, using 2 eq. of methyl acrylate as hydrogen acceptor and 5 mol% of $[RhCl(CO)(PPh_3)_2]$ (**11**) as catalyst, in a water/*N,N*-DMF solvent mixture. No conversion was observed in the absence of MW irradiation.

Primary alcohols, such as *n*-heptanol, *n*-tridecanol or 1,7-heptanediol were oxidized using a 2.5 mol% of $[RuCl_2(PPh_3)_2]$ (**12**) and 2 eq. of methyl vinyl ketone under solvent free conditions and MW heating at 120 °C for 15 min [34].

More recently, the efficient ruthenium-catalyzed C-3 reductive alkylation of 4-hydroxycoumarin by dehydrogenative oxidation of benzylic alcohols (Scheme 2) was reported [35]. Under optimized conditions (type of catalyst and solvent, activation method, reaction time, temperature and presence of base), using $[RuCl_2(PPh_3)_3]$ (12) as catalyst in *tert*-amyl alcohol under MW irradiation at 140 °C for 2 h, afforded good selectivity/conversion.

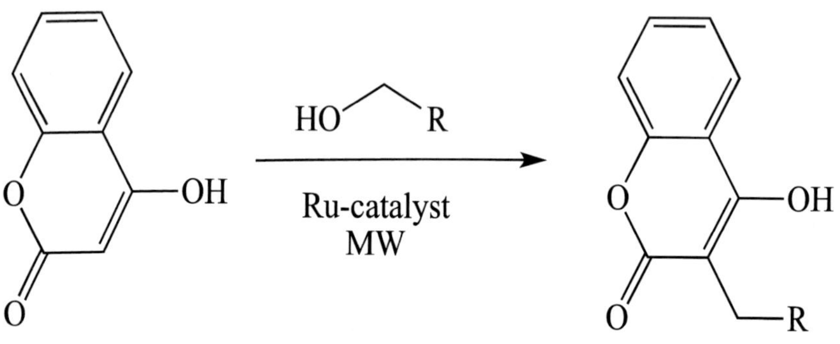

Scheme 2. MW-assisted selective C-3 alkylation of 4-hydroxycoumarin with benzyl alcohol catalyzed by 12.

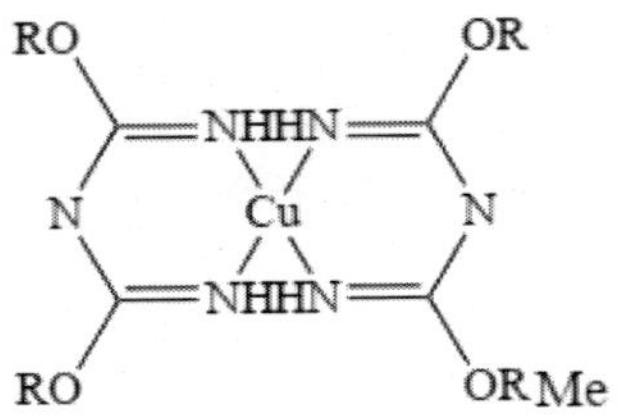

R = Me (**13**); R = Et (**14**)

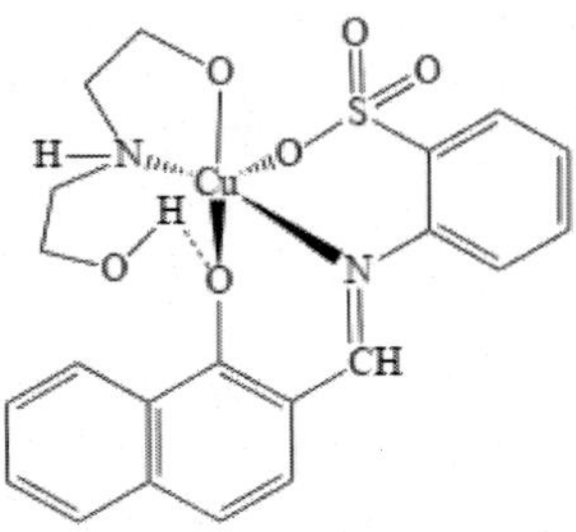

(**15**)

Figure 5. Copper(II) complexes [Cu{HN=C(OMe)NC(OR) = NH}$_2$] [R = Me (**13**) or Et (**14**)] and [Cu(H_2R)(HL)]·H_2O (**15**).

Several literature examples use copper(II) or Cu(II)/ 2,2,6,6-tetramethylpiperidyl-1-oxyl (TEMPO,) catalytic systems [26, 27, 30, 36-43]. For example, alkoxy-1,3,5-triazapentadien(e/ato) copper(II) complexes [Cu{HN=C(OMe)NC(OR)=NH}$_2$] [R = Me (**13**) or Et (**14**), Figure 5] allowed acetophenone yields up to 97%, TONs up to 485 and TOFs up to 1.2 x 10^3 h^{-1} [36] and the water-soluble copper(II) complex [Cu(H_2R)(HL)]·H_2O (**15**) [prepared by reaction of copper(II) nitrate hydrate with (*E*)-2-(((1-hydroxynaphthalen-2-yl)methylene)amino) benzenesulfonic acid (H_2L) and diethanolamine (H_3R) [30] (Figure 5)], led, in the presence of TEMPO, to acetophenone yields up to 85% in a remarkably short reaction time (15 min.), with the corresponding TOF value of 3.40 × 10^3 h^{-1} under low power MW irradiation.

Metal azolate/carboxylate (MAC) compounds [Cd(dmpzc)(DMF)(H_2O)] (**16**), [Pd(H_2dmpzc)$_2$$Cl_2$] (**17**), [Cu(Hdmpzc)$_2$] (**18**), [$Zn_4$O(dmpzc)$_3$]·Solv (**19**), and [$Co_4$O(dmpzc)$_3$]·Solv (**20**), isolated by coupling 3,5-dimethyl-1*H*-pyrazol-4-carboxylic acid (H_2dmpzc) to Cd(II), Pd(II), Cu(II), Zn(II), and Co(II) salts, were reported as catalysts towards alcohols oxidation [37]. **18**, a 3-D non-porous MAC framework, was remarkably active in the solvent-free microwave-assisted oxidation of 1-phenylethanol to acetophenone (yields up to 99% at 120 °C in only 0.5 h).

The same researchers reported the coupling of five, rigid or flexible, bis(pyrazolato)-based tectons with late transition metal ions allowing the isolation of eighteen coordination polymers (CPs) [38]. Two representative families of CPs, prepared with either rigid or flexible spacers, were tested as catalysts in the microwave-assisted solvent-free peroxidative oxidation of alcohols by *t*-BuOOH. The authors found that those CPs bearing the rigid spacer, concurrently possessing higher specific surface areas, were more active than the corresponding ones with the flexible spacer.

Moreover, the two copper(I)-containing CPs exhibited the highest efficiency, leading selectively to a maximum product yield of 92% (TON up to 1.5×10^3) in the oxidation of 1-phenylethanol. The favorable effect of MW irradiation was observed even when applying the low power of 5 W: for example, only 15% of product was obtained under the same conditions but using conventional heating (an oil-bath). As a further confirmation of this observation, after 15 h reaction, yields of 89 and 58% were obtained for MW and conventional heating, respectively, in the presence of the said copper(I) catalyst [38].

In a related study, temperature dependent synthesis of the dicopper(II) complex [Cu_2(L-κ*ONO*')$_2$(μ-4,4'-bipy)(DMF)$_2$] (**21**) and of the dicopper(II) based coordination polymer [Cu_2(μ-L-1κ*ONO*':2κ*O*)$_2$(μ-4,4'-bipy)]$_n$.nH_2O.nDMF (**22**) (Figure 6), derived from the acyclic Schiff base 2-[(2-hydroxy-3-methoxyphenyl)methylideneamino]benzenesulfonic acid (H_2L), was recently performed and **21** and **22** used as catalysts for both primary and secondary alcohols oxidation [39]. They exhibited good catalytic activity for the solvent- and additive-free MW-promoted peroxidative (with *tert*-butylhydroperoxide) oxidation of cyclohexanol (up to 85% yield of cyclohexanone) after 30 min of low power MW irradiation.

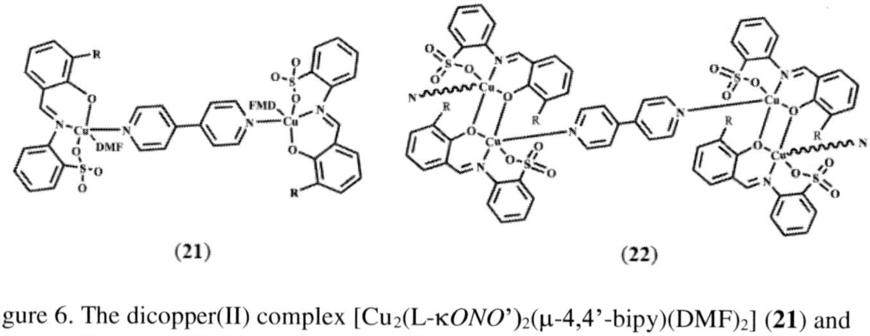

(**21**) (**22**)

Figure 6. The dicopper(II) complex [Cu_2(L-κ*ONO*')$_2$(μ-4,4'-bipy)(DMF)$_2$] (**21**) and dicopper(II) based CP [Cu_2(μ-L-1κ*ONO*':2κ*O*)$_2$(μ-4,4'-bipy)]$_n$.nH_2O.nDMF (**22**).

Previously, the same authors, by taking advantage of the coordinating ability of the sulfonate group of the acyclic Schiff base 2-[(2-hydroxy-3-methoxyphenyl) methylideneamino]benzenesulfonic acid H_2L, managed to synthesize two mononuclear copper complexes $[CuL(H_2O)_2]$ (**23**), $[CuL(2,2'\text{-bipy})].DMF.H_2O$ (**24**) and two diphenoxo-bridged dicopper compounds $[CuL(py)]_2$ (**25**) and $[CuL(EtOH)]_2.2H_2O$ (**26**) (Figure 7) [40] and to apply them as selective catalysts for the homogeneous peroxidative (with TBHP) oxidation of primary and secondary alcohols, under solvent- and additive-free conditions and under low power MW irradiation. A quantitative yield of acetophenone was obtained by oxidation of 1-phenylethanol with compound **25** (TOFs up to 7.6×10^3 h^{-1}) after 20 min of MW irradiation, whereas the oxidation of benzyl alcohol to benzaldehyde is less effective (TOF 992 h^{-1}). The selectivity of **25** to oxidize the alcohol relative to the ene function was demonstrated when using cinnamyl alcohol as substrate [40].

The catalytic systems **25** or **26**/THBP/MW are among the fastest and most effective copper catalyzed MW-enhanced, solvent- and additive-free oxidations of secondary alcohols. In fact, the used low power (10 W) MW irradiation provided a much more efficient synthetic method than conventional heating (open atmosphere or non-pressurized refluxing), allowing the attainment of similar or better yields in much shorter times (Figure 8): at 100 °C, the MW-assisted method with **25** leads to the almost quantitative oxidation of 1-phenylethanol to acetophenone in only 20 mins, whereas by the conventional method of heating, even at the higher temperature of 150 °C, the quantitative conversion was not yet approached after 3.5 h. Attempts to perform the oxidation at room temperature (without MW or conventional heating) failed, and the minimum required temperature was 50 °C. The overall temperature coefficient (in the 50–100 °C range) of these MW oxidations was ca. 1.2.

Other classes of copper effective heterogeneous catalysts for the solvent- and additive-free microwave-assisted peroxidative oxidation of alcohols are Cu(II) complexes with arylhydrazones of active methylene

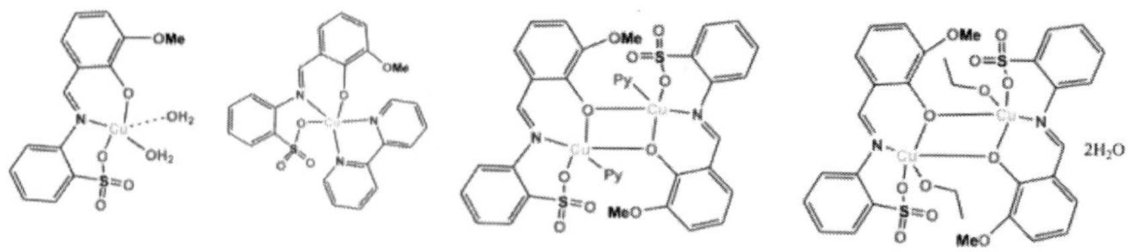

(23) (24) (25) (26)

Figure 7. Mononuclear copper complexes [CuL(H_2O)$_2$] (**23**) and [CuL(2,2'-bipy)].DMF.H_2O (**24**), and diphenoxo-bridged dicopper compounds [CuL(py)]$_2$ (**25**) and [CuL(EtOH)]$_2$.2H_2O (**26**) {H_2L = 2-[(2-hydroxy-3-methoxyphenyl) methylideneamino]benzenesulfonic acid}.

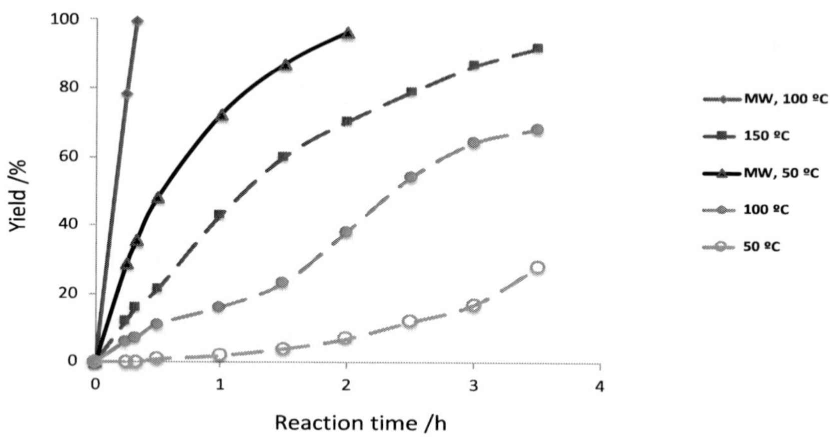

Figure 8. Influence of the temperature and reaction time on the yield of acetophenone obtained by MW-induced (–) or conventional (--) oxidation of 1-phenylethanol in the presence of **25**.

compounds (AHAMCs) ligands (e.g., [$Cu_4(H_2O)_{10}(1\kappa N{:}\kappa^2 O{:}\kappa O,2\kappa N{:}\kappa O\text{-}L^2)_2]\cdot 2H_2O$ (**27**) ($Na_2H_2L^2$ = sodium 5-(2-(4,4-dimethyl-2,6-dioxocyclohexylidene)hydrazinyl)-4-hydroxybenzene-1,3-disulfonate]) [41] or with arylhydrazone of ethyl 2-cyanoacetate or formazan ligands [42].

A good catalytic activity was observed for **27** towards oxidation of 1-phenylethanol, leading to yields of acetophenone up to 85.5% and TOF up to 1.90×10^3 h^{-1} (i.e., 4.75×10^2 h^{-1} per Cu atom) for a very low catalyst loading (catalyst / substrate molar ratio of 0.04%) after 1 h under MW irradiation [41]. Interestingly, a significant yield decrease to 58.1% was observed for the 1-phenylethanol oxidation in the presence of TEMPO. Moreover, this heterogeneous catalyst could be easily recovered and reused, at least for three consecutive cycles, maintaining 89% of the initial activity and a high selectivity.

The aquasoluble Cu(II) complexes $[Cu(1\kappa N,O^2{:}2\kappa O\text{-}HL^1)(X)]_2$ [X = CH_3OH (**28**) or $(CH_3)_2NCHO$ (**29**)] and $[Cu(\kappa N\text{-}HL^1)(en)_2]\cdot CH_3OH\cdot H_2O$ (**30**) prepared by reaction of copper(II) nitrate hydrate with the new ligand (*E*/*Z*)-4-(2-(1-cyano-2-ethoxy-2-oxoethylidene)hydrazinyl)-3-hydroxybenzoic acid (H_3L^1), in the presence (for **30**) or absence (for **28** and **29**) of ethylenediamine (Figure 10) [42] act as catalyst precursors for the solvent-free microwave assisted selective oxidation of primary or secondary alcohols and diols to the corresponding aldehydes, ketones and diketones, respectively, with yields in the 5–99% range (TONs up to 4.96×10^2) after 60 min of MW irradiation at 120 ºC.

The Cu(II) complex bearing one arylhydrazone of ethyl 2-cyanoacetate ligand, coordinated via metal-cyano group (**30**) evidenced the highest catalytic activity, due to the ability of the coordinated species to leave the metallic center and dispose free coordination sites, promoting the reaction with the alcohol substrates and/or oxidant. Again, efforts to increase the catalytic activities using TEMPO were unsuccessful, which end up to revealing an opposite desired effect [42].

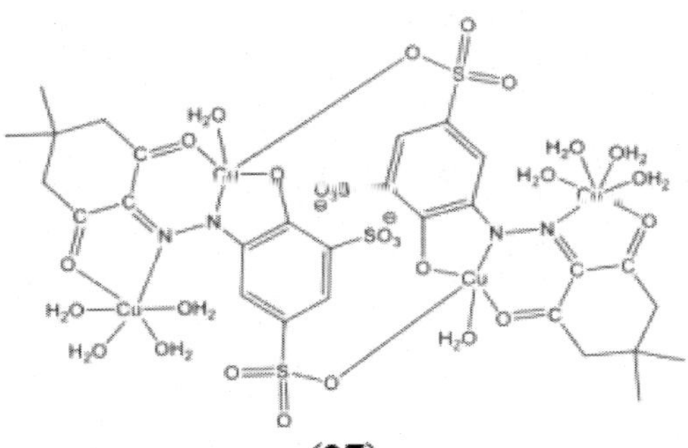

(27)

Figure 9. The copper tetramer $[Cu_4(H_2O)_{10}(1\kappa N{:}\kappa^2 O{:}\kappa O,2\kappa N{:}\kappa O\text{-}L^2)_2]\cdot 2H_2O$ (**27**) [$Na_2H_2L^2$ = sodium 5-(2-(4,4-dimethyl-2,6-dioxocyclohexylidene)hydrazinyl)-4-hydroxybenzene-1,3-disulfonate].

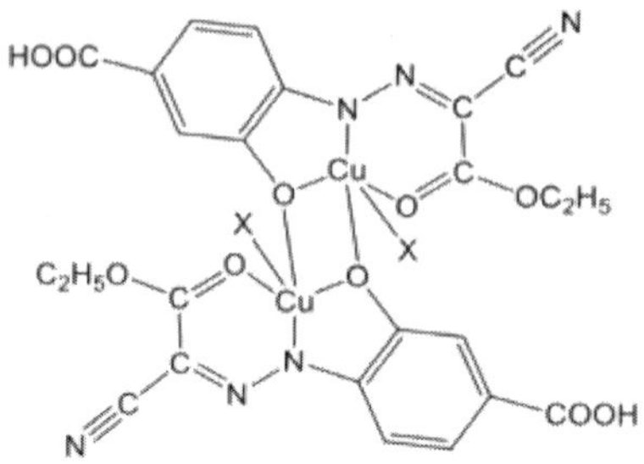

X = CH_3OH (**28**) or $(CH_3)_2NCHO$ (**29**)]

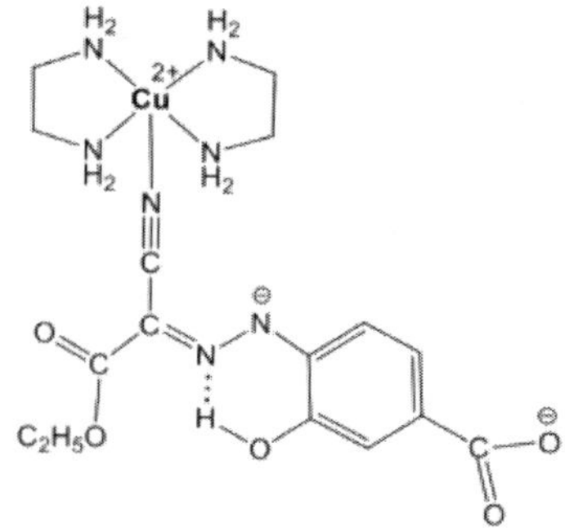

(**30**)

Figure 10. Cu(II) complexes $[Cu(1\kappa N,O^2{:}2\kappa O\text{-}HL^1)(X)]_2$ [X = CH_3OH (**28**) or $(CH_3)_2NCHO$ (**29**)] and $[Cu(\kappa N\text{-}HL^1)(en)_2]\cdot CH_3OH\cdot H_2O$ (**30**), {H_3L^1 = (*E*/*Z*)-4-(2-(1-cyano-2-ethoxy-2-oxoethylidene)hydrazinyl)-3-hydroxybenzoic acid}.

MW irradiation afforded an enhanced acetophenone yield with both precatalysts **28** and **30**: 94.0% (**28**) and 95.0% (**30**) yield was reached just after 30 min and full conversion within 1 h of reaction. In contrast, conventional (silicone bath) heat led at 30 min 52.5 and 53.3% for **28** and **30**, respectively, and at 1 h 97.6 and 94.4% for **28** and **30**, respectively. Furthermore, it was necessary to carry out the reaction for 3 h to obtain full alcohol conversion with the conventional heating mode. The authors pointed out the low MW irradiation power (less than 20 W) required making this synthetic protocol more desirable than the conventional one (requiring ca. 1500 - 2000 W to achieve a 120 °C reaction temperature).

Oxidation, with peroxide, of primary aromatic alcohols (benzyl and cinnamyl alcohol) end up being much less efficient than secondary ones (1-phenylethanol and benzhydrol). Despite the less steric hindrance of primary alcohols that should promote the reactivity of the substrate, lower conversions were obtained (Figure 11), conceivably due to the higher stability of the radical intermediate formed from secondary alcohols ($R_2C^{\cdot}$–OH> $RHC^{\cdot}$–OH) upon the H abstraction (radical mechanism).

Cyclooctanone was obtained in higher yields than cyclohexanone, in accord with the greater instability/reactivity of the 8-carbon member ring. Moreover, camphor is produced in higher yield from isoborneol than fenchaldehyde from fenchyl alcohol. The justification proposed by the authors relies on the difference of the steric hindrance of -OH group in the alcohol substrate.

The corresponding diketones of 1,2- and 1,4-cyclohexanediol were obtained with high selectivity, and again similar activities were observed for both catalyst precursors **28** and **30**. Vicinal diketones (51.5 and 56.0% for **28** and **30**, respectively) were formed in higher yields than 1,4-diketones (18.1 and 23.4% for 28 and 30, respectively). This tendency was expected in *vic*-diols since after one H· abstraction, the remaining H of the other –OH functional group has increased acidity (higher pK_a) and therefore became easier to be captured by radicals from the peroxide.

The 2D metal–organic framework (MOF) with pyridyl-based isophthalic acid $[\{Cu(L1)(DMF)\}.DMF.H_2O]_n$ (**31**) (H_2L1= 5-{(pyridin-4-ylmethyl)amino}isophthalic acid) [43] acts as heterogeneous polymeric solid catalyst for solvent-free microwave assisted peroxidative oxidation of primary and secondary alcohols. Under the above conditions, 81 or 98% yield of benzaldehyde or acetophenone and TOF of 811 or 983, respectively, are obtained in the absence of any additive. This copper MOF-based heterogeneous catalyst is of easy recovery and was reused in consecutive cycles without losing activity [43].

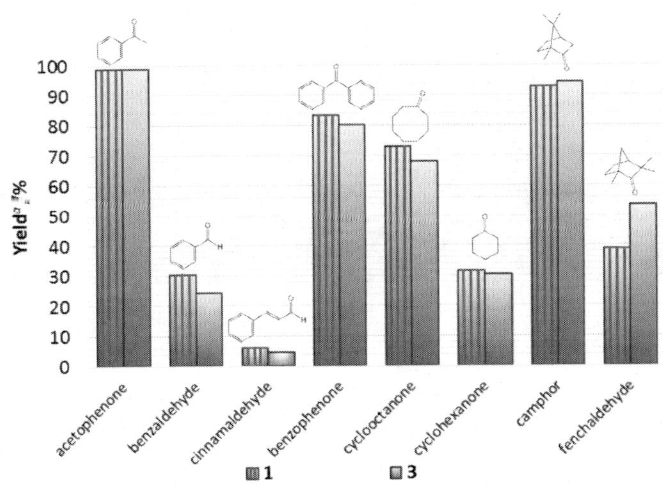

Figure 11. Conversions from MW-induced solvent-free oxidation of several alcohols using 28 (■) or 30 (■) as catalyst precursors.

A series of mixed-ligand dinuclear manganese(II) Schiff base complexes were disclosed as catalysts for the MW assisted oxidation of alcohols [31, 32]. Acetophenone yield of 81% is obtained using a maximum of 0.4% molar ratio of $[Mn(H_2L)\text{-}(py)_2]_2(NO_3)_2\cdot 2CH_3OH$ (**32**, H_2L = hydrazone Schiff base) catalyst relative to the substrate (1-phenylethanol) in the presence of TEMPO and in aqueous basic solution, under mild reaction conditions [32]. However, when compared with copper complexes, although equally selective, Mn complexes catalytic activity is significantly lower, leading to moderate alcohol conversions.

Catalytic industrial applications of nanoparticles (NPs) are a reality since long. Nanotechnology knowledge can be applied to improve catalytic organic transformations where the nanometer dimension of a catalyst can enhance its activity, due to its high specific surface area (surface area per mass). As a strategy for single-pot efficient oxidative functionalization of alcohols under mild conditions Au nanoparticles supported on Al_2O_3, Fe_2O_3, ZnO and TiO_2 have been tested in the oxidation of benzyl alcohol and methylbenzyl alcohol in the presence of *tert*-butylhydroperoxide under MW irradiation [44]. All materials were highly selective towards the formation of benzaldehyde or acetophenone, respectively. Addition of gold increased the alcohol conversion from 5% (for TiO_2) to 91% (for Au/TiO_2), which was the best result obtained. Au/TiO_2 recyclability was tested up to 10 cycles and catalytic activity remained at high level in the first 4 cycles. The loss of activity was due to a large increase in gold nanoparticle size and gold leaching after the 10th cycle.

Although heterogeneous catalysis by (supported) gold nanoparticles remains predominant, an explosive growth of homogeneous gold catalysis took place a few years ago and keeps increasing. Very recently, commercially available Au(I) or Au(III) compounds (Figure 12), namely, dichloro(2-pyridinecarboxylato)gold(III) (**33**), tetrabutylammonium tetrachloroaurato(III) (**34**), chlorotrimethylphosphinegold(I) (**35**), chlorotriphenylphosphinegold(I) (**36**) and 1,3-bis(2,6-diisopropylphenyl) imidazol-2-ylidenegold(I) chloride (**37**) supported on different carbon materials: activated carbon (AC), multi-walled carbon nanotubes (MWCNT) and carbon xerogel (CX), with three different surface treatments:

original (as purchased/prepared), oxidized with nitric acid (-ox), and oxidized with nitric acid and subsequently treated with NaOH (-ox-Na) were tested in the selective oxidation of methyl benzyl alcohol, cyclohexanol and 2-octanol with TBHP, under MW irradiation, to the corresponding aldehydes or ketones [45].

In general, better conversion results were obtained for the heterogenised complexes at the most efficient support, CNT-ox-Na. The highest activity for methyl benzyl alcohol oxidation was found for the **36**/CNT-ox-Na material, which could be recycled up to 7 cycles without activity or selectivity loss. Thus, functionalization is beneficial, as complexes supported on –ox materials are more active than when supported on pristine samples.

A promising perspective for catalysts recycling concerns the application of magnetically recoverable catalysts that can be readily collected by magnetic (external field) attraction avoiding the use of traditional isolation methods, such as filtration, extraction or centrifugation.

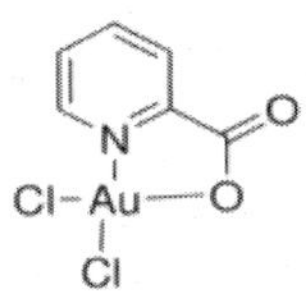

(33)

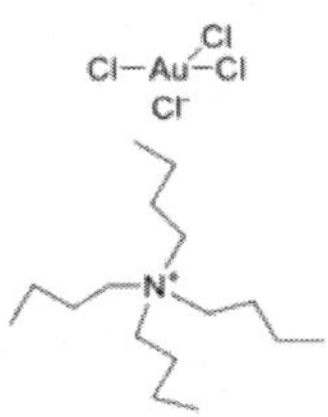

(34)

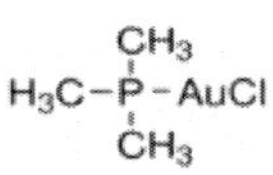

(35)

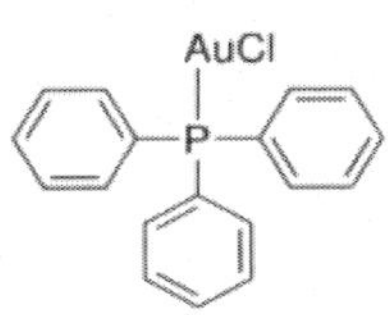

(36)

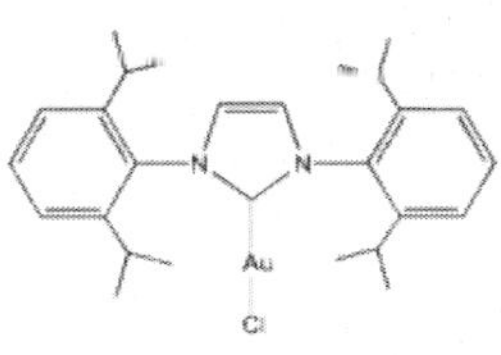

(37)

Figure 12. Commercially available Au(I) and Au(III) compounds **33 – 37**, catalysts for alcohol oxidation.

A type of monodispersed nanomaterials that has been widely explored are spinel ferrites, due their electronic, optical, electrical, magnetic and catalytic properties, which are different from their bulk counterparts. These ferrites, with the standard formula MFe_2O_4, where M is a divalent cation, offer more interesting catalytic properties compared to the corresponding single component metal oxides [46-51], thus having a huge potential use as heterogeneous catalysts towards various organic transformations. If the catalytically active species are non-magnetic, they can be immobilized on the surface of a magnetic carrier (e.g., Fe_2O_3 or Fe_3O_4) be applied for the oxidation of alcohols, and then be recovered by application of an external magnetic field. Therefore, such simple and easy catalyst separation and recycling introduces new synthetic procedures of significance in green chemistry.

In addition, ferrites or materials composed of ferrite powder have attracted a great interest for their strong MW absorption ability, contributing for better catalytic performances [52] and being used for e.g., MW-induced degradation of environmental organic contaminants [53].

First-row-transition-metal ferrite magnetic nanoparticles (NPs) MFe_2O_4 [M = Mn^{2+} (**38**), Fe^{2+} (**39**), Co^{2+} (**40**), Ni^{2+} (41), Cu^{2+} (**42**) or Zn^{2+} (**43**)], prepared by co-precipitation method [46], were used as catalysts for the microwave-assisted oxidation of various alcohols in a solvent-free medium. $MnFe_2O_4$ (**38**), $CoFe_2O_4$ (**40**) and $CuFe_2O_4$ (**42**) act as catalysts for the conversion of alcohols to the corresponding ketones or aldehydes with a yield range of 81 to 94% in 2 h at 120 °C using *t*-BuOOH [46]. These catalysts can be readily isolated from the reaction medium by using an external magnet and no significant loss of activity was observed when reused up to 10 consecutive runs. The advantages of **38-43** NPs catalytic systems in terms of efficiency and economy, is due to the used MW heating mode which generates heat evenly throughout the reactor, enhancing product purity, selectivity and yield, as well as being more energy efficient and economical in comparison to conventional heating methods.

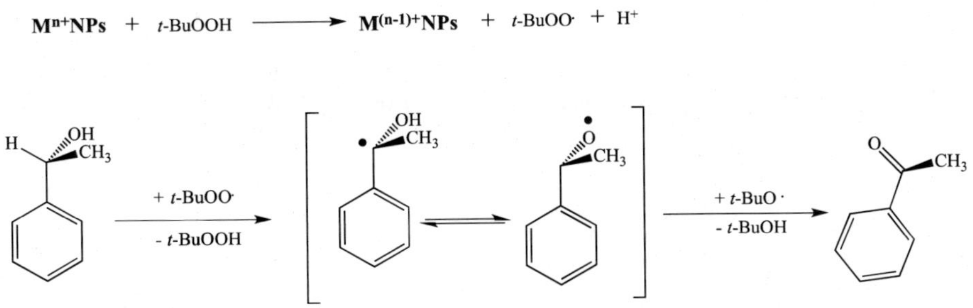

Scheme 3. Proposed radical mechanism for the model reaction, oxidation 1-phenylethanol to acetophenone, catalyzed by magnetic **39** NPs.

The oxidation of methyl benzyl alcohol using **39** NPs was performed in the presence of different oxidizing agents such as atmospheric oxygen, hydrogen peroxide and *tert*-butyl hydroperoxide. The highest activity was achieved with *t*-BuOOH, which proved once more to be more powerful than H_2O_2 towards alcohol oxidation [3]. The attempt to promote the catalytic performance by adding the organic radical 2,2,6,6-tetramethylpiperidin-1-yl)oxidanyl (TEMPO, 2.5% mol vs. alcohol substrate) was unsuccessful, resulting in an undesired inhibition effect. The addition of known C- or O-radical traps, Ph_2NH or $CBrCl_3$ resulted in a strong inhibition of product formation, indicating the radical reaction nature (Scheme 3) of the catalytic system using *t*-BuOOH [3, 42].

A series of alcohol substrates (benzyl and cinnamyl alcohol, as well as cyclo -pentanol, -hexanol, -heptanol and -octanol) was further evaluated by the same authors [46]. All substrates were oxidized to the corresponding aldehydes or ketones with very high selectivity (>99%). Secondary aromatic alcohols were oxidized in a larger extent than primary ones on account of the higher stability of the radical organic species formed upon hydrogen abstraction (R_2C·-OH vs. RHC·-OH), in spite of the lower steric hindrance for the latter. Note that the C = C double bond was not affected during the oxidation of cinnamyl alcohol to cinnamaldehyde. The tendency for lower conversion of cyclic aliphatic secondary alcohols in comparison with the aromatic ones can also be explained based on the stability of the radical intermediates in which the aromatic ring plays a crucial role (see above).

Among the cyclic aliphatic secondary alcohols, the cycloketone yield increases with the ring size: C5 < C6 < C7 < C8 [42].

It seems that more important than particle size/surface morphology is the nature of the divalent cation (M^{2+}) of MFe_3O_4 regarding the activity towards benzyl alcohol and 1-phenylethanol oxidation. The higher yields of ketone/aldehyde were obtained with the NPs that are constituted by Mn (**38**), Co (**40**) and Cu (**42**).

In a related work [54], EDTA functionalized magnetic NPs with general formula Fe_3O_4@EDTA-M^{2+} [M= Mn^{2+} (**44**), Fe^{2+} (**45**), Co^{2+} (**46**), Ni^{2+} (**47**), Cu^{2+} (**48**) or Zn^{2+} (**49**); EDTA = ethylenediamine tetraacetic acid] were prepared by a co-precipitation method and applied towards the microwave-induced oxidation of several alcohol substrates in solvent-free medium. This study demonstrates that **44** - **49** MNPs act as efficient catalysts, presenting high selectivity and yields up to 99% after 2 h of reaction at 110 °C using *t*-BuOOH as oxidant. Secondary alicyclic alcohols were oxidized with yields over than 95%, with cyclooctanol exhibiting a slightly higher conversion than the other alcohols with shorter carbon rings and a lower capacity to form radical intermediate structures. Moreover, camphor is produced in a higher yield from isoborneol (90%) than fenchaldehyde from fenchyl alcohol (67%) conceivably on account of steric effects.

The catalytic behavior of **44** MNPs was extended to the oxidation of two isomeric diols: 1,2- and 1,4-cyclohexanediols. The vicinal diketone (91%) was formed in higher yield than the 1,4-diketone (73%). This tendency was expected in *vic*-diols (see above) [42]. Excellence in selectivity was not achieved in diols oxidation (ca. 92%) due the formation of e.g., hydroxyl ketones or hydroxyoxeptanones (Baeyer-Villiger mechanism type) [42, 55, 56].

44-49 EDTA functionalized magnetite nanocatalysts constitute an optimization of the magnetic ferrites previously reported by the same authors [46] by limiting the aggregation effect between nanoparticles that would reduce the available surface area. Therefore, a higher specific area was obtained and more active sites per mass unit able to catalyze the oxidation reaction were obtained. Importantly, these improved features occur without loss of the magnetic properties. Therefore, the catalytic performance of EDTA functionalized materials allowed achieving full conversion of 1-phenylethanol, not accomplished using **38-43** MNPs [46].

A MW-promoted solvent-free peroxidative oxidation of benzyl alcohol to benzaldehyde catalyzed by low load of magnetic Ni-doped $MgFe_2O_4$ NPs was also successfully reported. The catalyst was reusable and the reaction time short [57].

The researchers' catalyst recycling strategy also includes the use of room temperature ionic liquids (RTIL) [53, 58] and supercritical fluids (e.g., supercritical CO_2, s-CO_2) [59], in some cases, with the advantage of enhancing the product yield and selectivity. RTIL exhibit several inherent benefic properties, such as low vapor pressures, high thermal and chemical stability and non-flammability [53]. The use of a room-temperature ionic liquid (RTIL) in MW-promoted synthesis, as a (co-) solvent and/or (co-)catalyst, is an increasingly exploited area since the ionic nature of ILs allows them to very efficiently absorb MW energy from the reaction media.

The role played by the combination of an IL and MW was explored in the activation of H_2O_2 in [hmim]Br (hmim = 3-methylimidazolium), used as catalyst and solvent for the metal-free chemoselective oxidation of several alcohols into their corresponding carbonyl compounds [60].

MW and ultrasound activation methods have been used in the oxidation of five- to eight-membered cyclanols in the presence of H_2O_2/H_2WO_4 and several hydrophobic ILs (e.g., ammonium based IL Aliquat 336) as co-catalysts [61]. Quantitative oxidation of cyclohexanol, after only 30 min at 90 °C, was achieved in the presence of [Aliquat][NTf_2] {Aliquat = *N*-methyl-*N*,*N*-dioctyloctan-1-ammonium; NTf_2 = bis(trifluoromethylsulfonyl) imide)}. No effect of the aromaticity of the IL cation was observed. In contrast, the IL anion size influenced the ketone yields what can be a key factor for the IL choice.

Examples of application of MW and ultrasound radiation are *i*) one-pot, tandem oxidation of cyclic alcohols to their respective lactones using potassium peroxy-monosulfate as oxidant and an IL as a solvent [62] or *ii*) cyclohexanol oxidation by Oxone®, catalyzed by TEMPO in the presence of tetrabutylammonium bromide in [bmim][BF_4] (bmim = 1-butyl-3-methylimidazolium). Ultrasounds and microwaves reduced the reaction time from 8 (using normal heating) to 5 h and 0.5 h, respectively, with similar yields of *ca.* 80%.

Supercritical CO_2 is the most used supercritical fluid. Its favorable characteristics such as stability, low toxicity and relatively low critical temperature allow most compounds to be extracted with little damage. Furthermore, the solubility of many extracted compounds in CO_2 depends on the pressure, what can impart selectivity. The selective conversion of benzyl alcohol to benzaldehyde in s-CO_2 was reported [59] with a 50% aldehyde increase for a pressure increment of 7%, due to the different partition of substrate and products in the organic and supercritical phases, thus affecting both the reaction rate and selectivity.

Further non-conventional technologies are now emerging for the selective oxidation of alcohols to carbonyl containing compounds [63]. One example is the environment-friendly synthetic technique, mechanochemistry (e.g., by ball milling) [64] that is now starting to be tested as solvent-free mechanochemical catalytical procedure.

CONCLUSION

From a sustainable viewpoint, the industrially very important alcohols' oxidation reactions using catalytic amounts of metal complexes in the presence of green oxidants has attracted much attention in recent years, with particular focus on enhancing selectivity, mechanistic aspects and on the improvement oxidant agents.

Several metal complexes, manly Cu-complexes are efficient catalysts for the oxidation of primary and secondary alcohols. The immobilization of metal catalysts on solid supports (e.g., carbon materials) is a good strategy to improve their catalytic activity and to facilitate the separation and catalyst recycle.

Although important developments have been achieved in oxidation catalysis with metal complexes toward the development of sustainable processes, namely by using unconventional conditions (e.g., microwave heating, supercritical fluid media, ionic liquids) there are still challenges to be addressed.

ACKNOWLEDGMENTS

The author is grateful to the Fundação para a Ciência e a Tecnologia (FCT), Portugal, (projects PTDC/QEQ-ERQ/1648/2014 and UID/QUI/00100/2013) for financial support.

REFERENCES

[1] Arends, I. W. C. E., Sheldon, R. A. *Modern Oxidation Methods*, Wiley-VCH Verlag GmbH & Co., 2010, pp. 147-185.

[2] Browne, W. R., de Boer, J. W., Pijper, D., Brinksma, J., Hage, R., Feringa, B. L. *Modern Oxidation Methods*, Wiley-VCH Verlag GmbH & Co. 2010, pp. 371-419.

[3] Kopylovich, M. N., Ribeiro, A. P. C., Alegria, E. C. B. A., Martins, N. M. R., Martins, L. M. D. R. S., Pombeiro, A. J. L. *Adv. Organomet. Chem.* 2015, *63*, 91–174.

[4] Sheldon, R. A., Arends, I. W. C. E., Dijksman, A. *Catal Today* 2000, *57*, 157–166.

[5] Sheldon, R. A., Arends, I. W. C. E., Ten Brink, G-J., Dijksman, A. *Acc Chem Res.* 2002, *35*, 774–781.

[6] Mallat, T., Baiker, A. *Chem. Rev.* 2004, *104*, 3037–3058.

[7] Mizuno, N. *Modern heterogeneous oxidation catalysis*. Wiley-VCH Verlag GmbH & Co. KGaA, Weinheim, 2009.

[8] De la Hoz, A., Loupy A. (Eds) *Microwaves in Organic Synthesis* 3rd ed., Wiley-VCH, Weinheim, 2013.

[9] Leadbeater, N. E. (Ed.) *Microwave Heating as a Tool for Sustainable Chemistry*, CRC Press, Boca Raton, FL, 2010.

[10] Polshettiwar, V., Varma R. S. (Eds.) *Aqueous Microwave Assisted Chemistry, Synthesis and Catalysis*, RSC, Cambridge, 2010.

[11] Tanaka, K. (Ed.) *Solvent-free Organic Synthesis*, Wiley-VCH, Weinheim, 2003.

[12] Suslick, K. S., Price, G. J. *Ann Rev Mater Sci*. 1999, *29*, 295–326.

[13] Matuszek, K., Zawadzk, P., Czardybon, W., Chrobok, A. *New J Chem*. 2014, *38*, 237–241.

[14] Chatel, G., Monnier, C., Kardos, N., Voiro, C., Andrioletti, B., Draye, M. *Appl Catal A: Gen*. 2014, *478*, 157–164.

[15] Han, X., Poliakoff, M. *Chem Soc Rev*. 2012, *41*, 1428–1436.

[16] Mercadante, M. A., Leadbeater, N. E. *Green Process. Synth*. 2012, *1*, 499-507.

[17] Sutradhar, M., Martins, L. M. D. R. S., Guedes da Silva, M. F. C., Pombeiro, A. J. L. *Coord. Chem. Rev*. 2015, *301-302*, 200-239.

[18] Conte, V., Coletti, A., Floris, B., Licini, G., Zonta, C. *Coord. Chem. Rev*. 2011, *255*, 2165 –2177.

[19] Licini, G., Conte, V., Coletti, A., Mba, M., Zonta, C. *Coord. Chem. Rev*. 2011, *255*, 2345–2357.

[20] Sutradhar, M., Martins, L. M. D. R. S., Carabineiro, S. A. C., Guedes da Silva, M. F. C., Buijnsters, J. G., Figueiredo, J. L., Pombeiro, A. J. L. *ChemCatChem*, 2016, *8*, 2254-2266.

[21] Sutradhar, M., Martins, L. M. D. R. S., Guedes da Silva, M. F. C., Pombeiro, A. J. L. *Appl. Catal. A: Gen*. 2015, *493*, 50-57.

[22] Martins, L. M. D. R. S., Ribeiro, A. P. C., Carabineiro, S. A. C., Figueiredo, J. L., Pombeiro, A. J. L. *Dalton Trans*., 2016, *45*, 6816 – 6819.

[23] Martins, L. M. D. R. S., Almeida, M. P., Carabineiro, S. A. C., Figueiredo, J. L., Pombeiro, A. J. L. *ChemCatChem*, 2013, 5, 3847.

[24] Ribeiro, A. P. C., Martins, L. M. D. R. S., Carabineiro, S. A. C., Buijnsters, J. G., Figueiredo, J. L., Pombeiro, A. J. L. *ChemCatChem*, 2018, *10*, 1821-1828.

[25] Karmakar, A., Martins, L. M. D. R. S., Guedes da Silva, M. F. C., Hazra, S., Pombeiro, A. J. L. *Cat. Lett*., 2015, 1-11.

[26] Mahmudov, K. T., Kopylovich, M. N., Sabbatini, A., Drew, M. G. B., Martins, L. M. D. R. S., Pettinari C., Pombeiro A. J. L. *Inorg Chem.* 2014, *53*, 9946–9958.

[27] Nasani, R., Saha, M., Mobin, S. M., Martins L. M. D. R. S., Pombeiro A. J. L., Kirillov, A. M., Mukhopadhyay, S. *Dalton Trans.* 2014, *43*, 9944–9954.

[28] Kumpulainen, E. T. T., Koskinen A. M. P. *Chem Eur J.* 2009, *15*, 10901–10911.

[29] Hoover, J. M., Ryland, B. L., Stahl, S. S. *ACS Catal.* 2013, *3*, 2599–2605.

[30] Sabbatini, A., Martins, L. M. D. R. S., Mahmudov K. T. *Catal Comm.* 2014, *48*, 69–72.

[31] Alexandru, M., Cazacu, M., Arvinte, A. *Eur J Inorg Chem.* 2014, 120–131.

[32] Sutradhar, M., Martins, L. M. D. R. S., Guedes da Silva, M. F. C., Alegria, E. C. B. A., Liu, C-M., Pombeiro, A. J. L. *Dalton Trans.* 2014, *43*, 3966–3977.

[33] Ciriminna, R., Pagliaro, M. *Org Proc Res Dev.* 2009, *14*, 245–251.

[34] Takahashi, M., Oshima, K., Matsubara, S. *Tetrahedron Lett.* 2003, *44*, 9201–9203.

[35] Montagut-Romans, A., Boulven, M., Lemaire, M., Popowycz, F. *New J Chem.* 2014, *38*, 1794–1801.

[36] Figiel, P. J., Kopylovich, M. N., Lasri, J., Guedes da Silva, M. F. C., Frausto da Silva, J. J. R., Pombeiro, A. J. L. *Chem Commun.,* 2010, *46*, 2766–2768.

[37] Tăbăcaru, A., Xhaferaj, N., Martins, L. M. D. R. S., Alegria, E. C. B. A., Chay, R. S., Giacobbe, C., Domasevitch, K. V., Pombeiro, A. J. L., Galli, S., Pettinari, C. *Inorg. Chem.*, 2016, *55*, 5804-5817.

[38] Timokhin, I., Pettinari, C., Marchetti, F., Pettinari, R., Condello, F., Galli, S., Alegria, E. C. B. A., Martins, L. M. D. R. S., Pombeiro, A. J. L. *Cryst. Growth Des.*, 2015, *15*, 2303-2317.

[39] Hazra, S., Martins, L. M. D. R. S., Guedes da Silva, M. F. C., Pombeiro, A. J. L. *Inorg. Chim. Acta*, 2017, *455*, 549-556.

[40] Hazra, S., Martins, L. M. D. R. S., Guedes da Silva, M. F. C., Pombeiro, A. J. L. *RSC Adv.*, 2015, *5*, 90079–90088.

[41] Mahmudov, K. T., Sutradhar, M., Guedes da Silva, M. F. C., Martins L. M. D. R. S., Ribera, A., Nunes, A. V. M., Gahramanova, S. I., Marchetti, F., Pombeiro, A. J. L. *RSC Adv.,* 2015, *5*, 25979–25987.

[42] Martins, N. M. R., Mahmudov, K. T., Guedes da Silva, M. F. C., Martins, L. M. D. R. S., Pombeiro, A. J. L. *New J. Chem.*, 2016, *40*, 10071-10083.

[43] Karmakar, A., Martins, L. M. D. R. S., Hazra, S., Guedes da Silva, M. F. C., Pombeiro, A. J. L. *Cryst. Growth Des.*, 2016, *16*, 1837-1849.

[44] Martins, L. M. D. R. S., Carabineiro, S. A. C., Wang, J., Rocha, B. G. M., Maldonado-Hódar, F. J., Pombeiro, A. J. L. *ChemCatChem*, 2017, *9*, 1211-1221.

[45] Carabineiro, S. A. C., Martins, L. M. D. R. S., Pombeiro, A. J. L., Figueiredo, J. L. *ChemCatChem*, 2018, *10*, 1804-1813.

[46] Martins, N. M. R., Martins, L. M. D. R. S., Amorim, C. O., Amaral, V. S., Pombeiro, A. J. L. *Catalysts*, 2017, *7*, 222-239.

[47] Barbero, B. P., Gamboa, J. A., Cadús, L. E. *Appl. Catal. B: Env* 2006, *65*, 21-30.

[48] Xiao, S. H., Jiang, W. F., Li, L. Y., Li, X. J. *Mater Chem. Phys.* 2007, *106*, 82-87.

[49] Menini, L., Pereira, M. C., Ferreira, A. C., Fabris, J. D., Gusevskaya, E. V. *Appl. Catal. A: Gen* 2011, *392*, 151-157.

[50] Su, M., He, C., Sharma, V. K., Abou Asi, M., Xia, D., Li, X.-z., Deng, H., Xiong, Y. *J. Hazard. Mat.* 2012, *211–212*, 95-103.

[51] Karami, B., Hoseini, S. J., Nikoseresht, S., Khodabakhshi, S. *Chin. Chem. Let.* 2012, *23*, 173-176.

[52] El Ghandoor, H. E., Zidan, H. M., Khalil, M. M. H., Ismail, M. I. M. *Int. J. Electrochem. Sci.* 2012, 5734-5745.

[53] Polshettiwar, V., Varma, R. S. *Green Chem.* 2010, *12*, 743-754.

[54] Martins, N. M. R., Martins, L. M. D. R. S., Amorim, C. O., Amaral, V. S., Pombeiro, A. J. L. *Catalysts*, 2017, *7*, 335-354.

[55] ten Brink, G. J., Arends, I. W. C. E., Sheldon, R. A. *Chemical Reviews* 2004, *104*, 4105-4124.

[56] Vafaeezadeh, M., Hashemi, M. *Catal. Comm.* 2014, *43*, 169-172.
[57] Yousefi, M., Kamel, M. H., Kar, A., Attar Kar, M. K. *Main Group Chem.* 2013, *12*, 177–184.
[58] Kumar, R., Sharma, N., Sharma, A., Sinha, A. K. *Mol Divers.* 2011, *15*, 687–695.
[59] Caravati, M., Grunwaldt, J. D., Baiker, A. *Appl Catal A Gen*, 2006, *50*, 298–303.
[60] Shirini, F., Abedini, M., Atghia, S. V. *Green Solvents II Properties and Applications of Ionic Liquids,* Ch. 14 Springer, London, 2012.
[61] Chatel, G., Monnier, C., Kardos, N., Voiro, C., Andrioletti, B., Draye, M. *Appl Catal A: Gen*. 2014, *478*, 157–164.
[62] Matuszek, K., Zawadzk, P., Czardybon, W., Chrobok, A. *New J Chem.* 2014, 38, 237–241.
[63] Phillips, A. M. F., Pombeiro, A. J. L., Kopylovich, M. N. *ChemCatChem*, 2017, *9*, 217–246.
[64] Fontolan, E., Alegria, E. C. B. A., Ribeiro, A. P. C., Kopylovich, M. N., Bertani, R., Pombeiro A. J. L. *Inorg. Chim. Acta*, 2017, *455*, 653-658.

In: Alcohol Oxidation
Editor: Steffan Paterson
ISBN: 978-1-53614-604-2

Chapter 2

MULTICATALYTIC ENANTIOSELECTIVE CASCADE REACTIONS INVOLVING ALCOHOL OXIDATION

Ana Maria Faisca Phillips*[*], *Maximilian N. Kopylovich and Armando J. L. Pombeiro
Centro de Química Estrutural, Instituto Superior Técnico,
Universidade de Lisboa, Lisbon, Portugal

ABSTRACT

Cascade reactions allow the elaboration of complex molecules from simple starting materials in a very efficient way. Several bonds can be made in one-pot procedures without the need for intermediates to be isolated. In this chapter we examine how simple alcohols have been used in this manner as feedstocks for the preparation of a variety of enantioenriched substances.

Keywords: domino reactions, oxidants, hydrogen borrowing, redox-relay, synergistic catalysis, multifunctional catalysts, asymmetric synthesis

---

[*] Corresponding Author Email: anafaiscaphillips@tecnico.ulisboa.pt.

1. Introduction

The search for sustainable methods of production, based on simple, readily available starting materials, which make efficient use of energy sources and are atom-economical, while producing at the same time minimum amounts of waste to protect the environment, are a major goal pursued by synthetic chemists working in academia and industry today [1-3]. Cascade reactions are a possible solution [4-6]. In a cascade reaction, all the synthetic steps necessary to convert a substance into a final target take place in a single vessel. Hence, cascade reactions are commonly referred to as "one-pot" reactions. They are different from the classical "stop and go" methodologies, which involve isolation and in most cases also purification of intermediates, to ensure that each step proceeds smoothly. This is time-consuming, expensive, generates extra waste and even material losses. In cascade reactions, the product of the first reaction is used *in situ* as the substrate for the next and it so continues thereafter, with each new product becoming the starting material for the following step (Figure 1). Cascade reactions can be further differentiated into domino and sequential reactions. In the first case, all the reactants, reagents and catalysts are introduced at the start and all the ensuing steps take place under the same conditions. In sequential processes, which also take place in "one pot," after one reaction or a domino process, more reagents or catalysts may be added and the temperature, pressure and other conditions, e.g., pH, may be changed. At the end, the final product is isolated. This "one-pot multi-stage approach" may also be suitable for adaptation to industrial flow processes.

Although the concept of a cascade reaction is very appealing, it is not necessarily easy to put into practice, particularly when there are a few functional groups present in the intervening molecules or if more than one catalyst is required. Indeed, finding compatible reagents and reaction conditions is one of the major challenges to overcome. For example, combining enzymes with metal catalysts can be particularly difficult, since the metal ions can deactivate the proteins by coordination and vice-versa [7]. Compartmentalization strategies have been used in some cases with success, e.g., with membranes or polyvinyl alcohol cryogels [8]. In addition, enzymes are usually only active in water, a medium that is frequently incompatible with metal catalysts. However, despite the potential problems that may be found along the way, the development of cascade reactions is a very active field of research.

When more than one catalyst is used, the reaction mechanisms can differ, depending on which substrates are being activated (Figure 2). When one of the cascade cycles is promoted by one catalyst and the next by a different one, the process is referred to as a "relay process." If two catalysts act in the same cycle, there are two possibilities: the catalysts can either activate the same substrate jointly, in which case there is "double activation" or they can activate two different substrates separately, which then react with each other. In this case there is "cooperative" or "synergistic catalysis." Bifunctional activation occurs in many reactions promoted by organocatalysts, e.g., by chiral thioureas or chiral phosphoric acids and in this case one catalyst is capable of activating at the same time two substrates, or more, bringing them together in a specific orientation with respect to each other, facilitating the reaction in more than one way.

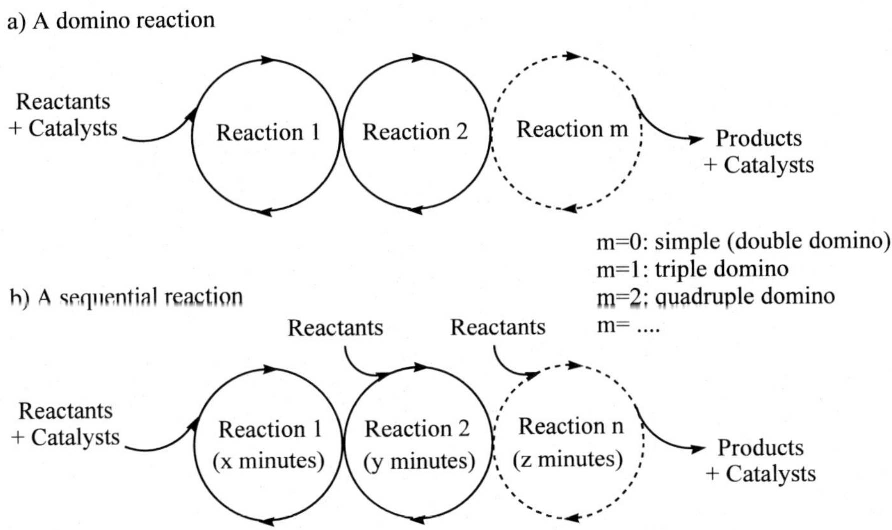

Figure 1. Types of "one-pot" cascade reactions.

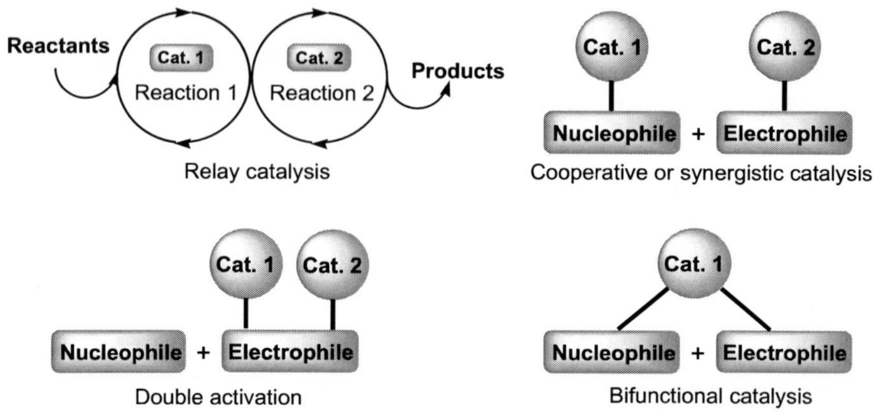

Figure 2. Types of substrate/reagent activation possible in multifunctional catalysis.

Alcohols are amongst the most used starting materials for synthesis. The lower alcohols are obtained mostly by well-established methods, e.g., acid-catalyzed hydration reactions, from ethylene, propene, or butenes, derived from naphtha (crude gasoline) or from synthesis gas, in the case of methanol [9]. Ethanol is nowadays produced more and more from fermentation processes of renewable resources such as plant-derived biomass, e.g., starches from corn, molasses, and grains from cereals, sorghum and barley, due to concerns over safeguarding existing fuel resources, in this society in which the demand for energy and chemicals is always rising [10]. It is estimated that in 2016 the global industrial alcohol market size had a value of USD 105.89 billion [11]. The fact that in some countries the blend of alcohol (ethanol) with gasoline is a requirement for fuel, is a driving force for increased production. Nevertheless, the lower alcohols are cheap raw materials, abundant, generally non-toxic and can be easily stored, transported and dissolved [12]. They can also be readily and selectively oxidized to aldehydes, ketones or acids, depending on their structure and on the nature of the oxidizing agent selected. These oxidation products are common chemical intermediates and oxidation from alcohols is indeed the major way in which they are produced in the laboratory.

Using alcohols as starting materials for a synthetic sequence brings several advantages [12]: they are more easily accessible and more stable; the aldehydes may polymerize or oligomerize easily under certain reaction conditions; some of their reactions may be difficult to control, e.g., the reactions of α-ketoaldehydes, due to their high reactivity [13]; alcohols are also less volatile than the aldehydes, which are susceptible to air oxidation and more toxic. Some are even lachrymatory. Hence it is very appealing to use alcohols directly as substrates for cascade reactions, in which the oxidation product is trapped and converted *in situ* into the next product as it is formed. Carbonyl compounds are more useful intermediates, since they display a wider range of reactivities, being able to undergo nucleophilic addition reactions as well as to be the nucleophile themselves, *via* the enolates.

This type of methodology was first reported by Huang and co-workers in 1987, for the synthesis of the amnesia reversal drug carbon-14-labelled CI-933 [14]. It was later further developed by Taylor and co-workers, who actually coined it "tandem oxidative processes" i.e., TOP [15, 16]. Williams and co-workers were the first to combine metal and enzyme catalysis in the same cascade process, in 1996, achieving in this way, in the following year, the resolution of alcohols described in section 5 [17]. In 2012 Rueping and co-workers showed that organocatalysts can also be used in TOP processes, and combined a metal-catalyzed oxidation of allylic alcohols with asymmetric iminium catalysis to obtain a variety of chiral products, e.g., formylcyclopropanes and 2-hydroxychromenones, with very high enantiomeric excesses (*ees*) [18]. This process is discussed in section 2.

The choice of oxidizing agent is crucial in cascade reactions, because although many reagents are available for alcohol oxidation [19, 20], only some are compatible with each process [16, 21]. The oxidizing agent must not react with other functional groups present in the reacting molecules, with any additional reagents used, or with the catalysts. It must not over oxidize the alcohol to the corresponding acid or even to the ester, unless the acid is the desired intermediate. Any competing reactions will lower yields and call for additional purification measures at the end. Figure 3 shows the structures of the oxidizing agents used in recent years in enantioselective cascade reactions initiated by alcohol oxidation.

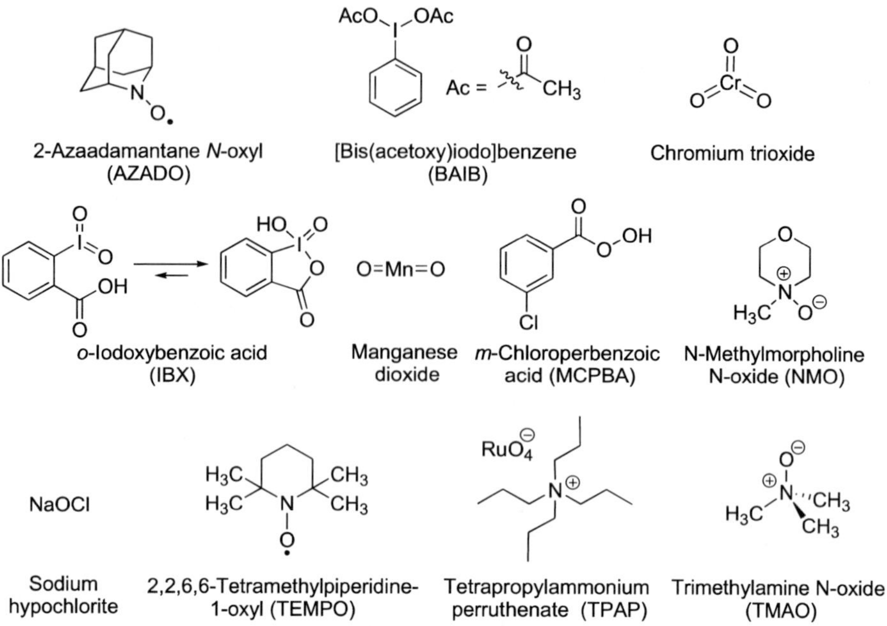

Figure 3. Oxidants used in enantioselective cascade reactions.

Enantioselective reactions bring additional constraints [12]. For example, whereas oxidations often occur at or above room temperature, enantioselective reactions may proceed well at moderate room temperatures but many are favoured by low temperatures. Chiral metal catalysts are sensitive to species present in the reaction medium with coordinating ability, which could be a problem when organocatalysts are used at the same time, since they frequently have heteroatoms with lone pairs, e.g., chiral amines or thioureas. This problem can be avoided with a sequential approach, which can also be employed when there are temperature issues. To avoid over oxidation of an aldehyde to an acid, the rate of the reaction promoted by a chiral organocatalyst used in a subsequent step has to be high. The undesirable formation of the acid will not only lower the yield of product which may be obtained, but it has also been observed in some cases that the presence of small amounts of acid can have a detrimental effect on the reactivity and enantiomeric excess (the *ee*) of some organocatalyzed reactions of aldehydes.

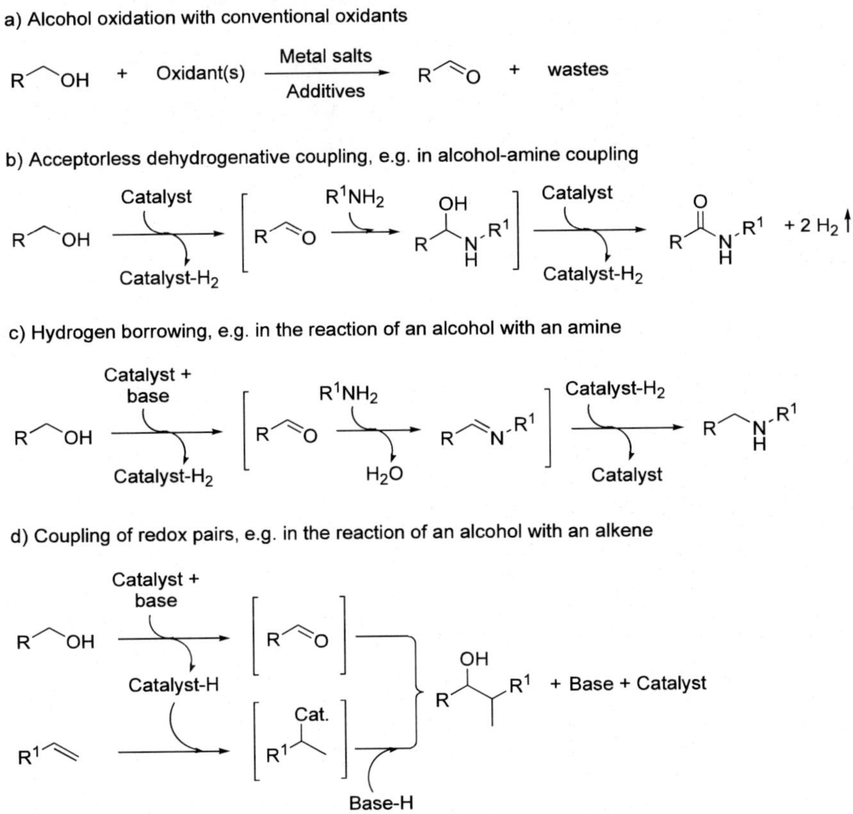

Figure 4. Methods to convert alcohols into value-added products.

However, despite the potential handicaps, many organocatalysts are very robust and are highly compatible with many reaction conditions. This, together with the fact that they can often promote reactions through several different activation modes, has allowed the development of very interesting and useful processes, in which many bond-breaking and bond-making steps take place in one pot in a highly stereoselective manner, leading to the formation in some cases of complex structures with many chiral centres.

Besides these obvious strategies, in recent years other related methods have been developed to convert alcohols into value-added products, which do not require the use of conventional oxidants: acceptorless dehydrogenative coupling, hydrogen borrowing and coupling of redox pairs (Figure 4) [22, 23]. In each case, mechanistically there is initial conversion of the alcohol to an aldehyde, which is used *in situ* for further reactions. Since direct oxidation methodologies can lead to the production of large amounts of waste, arising not only from the stoichiometric oxidant, but also from cocatalysts and additives, these variations are thought of as "greener" methodologies. The amount of waste generated is *a priori* less, because besides the substrates, only substoichiometric amounts of catalyst are required.

In the acceptorless dehydrogenative coupling of alcohols, a catalyst liberates hydrogen from the reactants or intermediates, leading to the formation of a coupled product with H_2 evolution [23, 24]. In some variations, water elimination occurs too. In the cases in which the catalysts back-hydrogenate an intermediate using the hydrogen removed from the starting compounds, the overall process is redox neutral, and is called "hydrogen borrowing or autotransfer" [25-27]. Finally, there are also reactions in which a redox pair, e.g., an alcohol and an alkene, are coupled, when intermediate nucleophiles and electrophiles are produced [28]. C-N and even C-C bonds can be made in this way in a very atom-economical fashion too.

Cascade rearrangements of propargylic alcohols and their derivatives are also examples of one-pot multistep processes which can be used for the synthesis of a diverse range of substances, e.g., enones, carbocycles and heterocycles [29].

The number of enantioselective processes developed with all these different methods is still small, but these are fast growing areas. In this chapter we focus on cascade reactions promoted by combinations of two catalysts, i.e., metals, organocatalysts and enzymes, or mixtures thereof, but not reactions catalyzed by enzymes only. For the later we refer the reader to some interesting reviews published recently [30-33]. Special hybrid catalysts, particularly designed so as to have distinct sites capable of activating molecules in more than one way, are also covered. Substantial savings in reaction time, effort, waste production, production costs and energy could be achieved with the cascade methodologies described in this chapter.

2. CASCADE REACTIONS PROMOTED BY ORGANOCATALYST-TRANSITION METAL COMBINATIONS

2.1. Methods Relying on Conventional Alcohol Oxidation

In the cascade reactions described in this section, all the reagents and catalysts were added to the reaction vessel at the start. The first enantioselective cascade process developed, which started with an alcohol oxidation, was reported by Kudo and co-workers in 2010 [34]. α-Oxyaminated aldehydes were prepared directly from primary alcohols *via* a domino process, consisting of oxidation/α-oxyamination. After reaction work-up, the aldehydes produced were subjected to borohydride reduction into the more stable alcohols before chromatography (Figure 5). Good to high yields of α-oxyaminated alcohols were obtained in very high *ees*. The alcohol to aldehyde oxidation was possible with a copper chloride/TEMPO (2,2,6,6-tetramethylpiperidine-1-oxyl) mixture and molecular oxygen as the terminal oxidant. The addition of copper chloride in substoichiometric amounts allows the use of TEMPO as catalyst for this reaction. The reaction mechanism probably involves the formation of a piperidinyloxyl copper(II) complex [35].

For this cascade, it was found that an additional ligand was necessary for copper, 2,2′-bipyridine, in the absence of which no oxidation took place. It was also found that the use of water as co-solvent improved the yields and *ees*. In its absence, over-oxidation to the carboxylic acid was observed. The aldehyde α-derivatization was possible because the reaction was performed in the presence of resin-supported peptide catalyst **2**, bearing a terminal proline residue, presumably *via* enamine catalysis. Seven recycling runs were possible without any deterioration in yields or enantioselectivity.

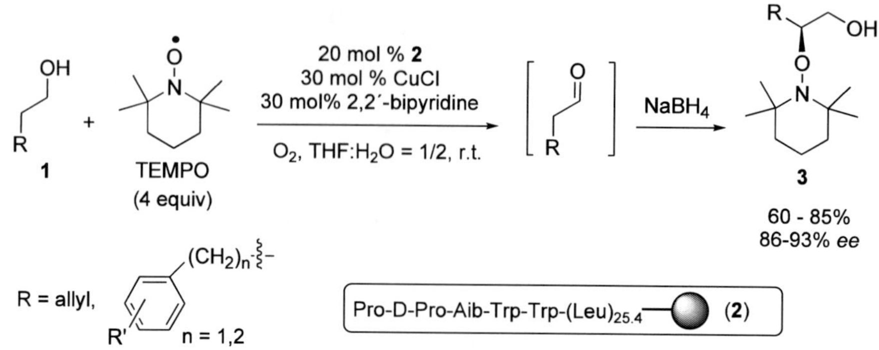

Figure 5. Oxidation/α-oxyamination cascade reaction of primary alcohols with TEMPO mediated by a resin-supported peptide.

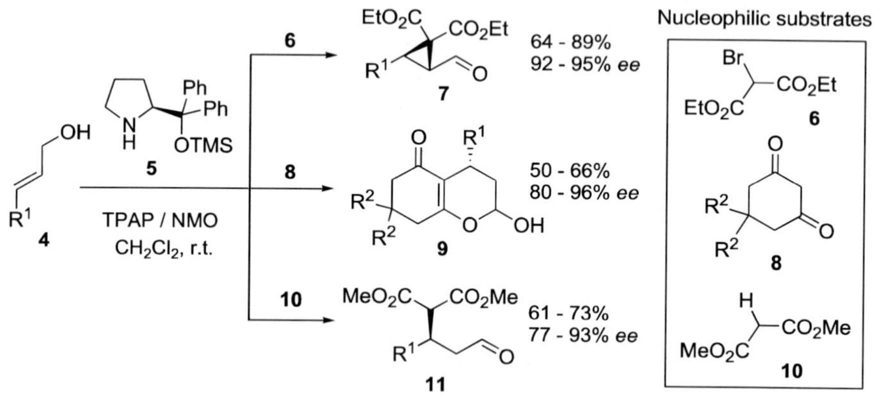

Figure 6. Domino reactions of allylic alcohols combining oxidation-iminium activation.

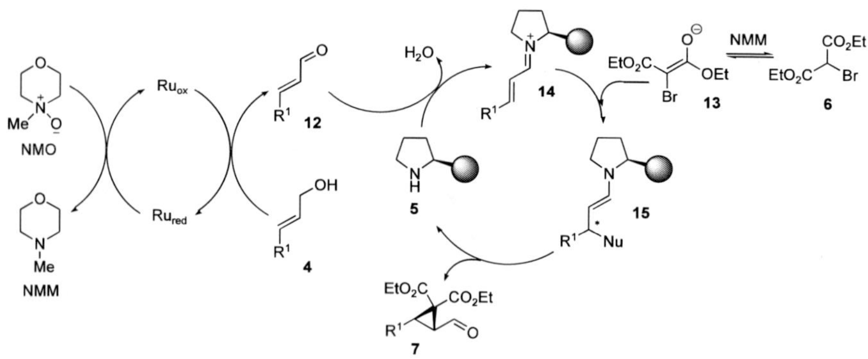

Figure 7. Mechanism proposed for the domino oxidation-cyclopropanation.

In 2012, Rueping and co-workers showed that the TPAP/NMO system could be combined with chiral amine catalysis. Tetrapropylammonium perruthenate (TPAP) is stable in organic media and it is a good oxidant for alcohols, e.g., **4** [18]. It is normally employed in catalytic quantities with *N*-methylmorpholine *N*-oxide (NMO) which acts as a secondary oxidant [36]. The TPAP/NMO system is very selective for alcohols in the presence of other functional groups [19]. It is air stable, non-volatile and soluble, and its compatibility with diphenylprolinol TMS-ether **5** allowed the development of a number of domino processes, which proceeded well at room temperature. The authors proposed that the *N*-methylmorpholine (NMM) produced from NMO in the alcohol to aldehyde oxidation, could act as a base to activate the nucleophilic substrates for further reactions, at the same time that the organocatalyst formed an iminium ion with the aldehyde generated *in situ*, which could subsequently undergo nucleophilic attack. This type of reactivity was used to produce several types of molecules. For example, the reaction of allylic alcohols **4** with diethyl bromomalonate (**6**) afforded formyl cyclopropanes **7** in high yields and excellent *ees* by an enantioselective oxidation/Michael addition/intramolecular alkylation domino (Figure 6). Similarly, when a 1,3-diketone **8** was selected as reaction partner, chromenones **9** could be produced by oxidation/1,4-addition/intramolecular cyclization (Figure 6). Malonates **10** could also be coupled to allylic alcohols *via* the asymmetric oxidative iminium cascade to provide products **11** with up to 93% *ee.* These were the first examples reported of enantioselective homogeneous relay cascade reactions initiated by alcohol oxidation. The mechanism proposed for these reactions is shown in Figure 7.

The oxidation/asymmetric cyclopropanation cascade of allylic alcohols and bromomalonate shown above could also be achieved with a different oxidant combined with the same chiral amine (**5**), as reported by the Rueping group around the same time [37]. In this case manganese dioxide was used as a heterogeneous oxidant and triethylamine as base, to provide the required nucleophile activation. No oxidation of the amine was observed under these conditions and the oxidant was removed by filtration. However, 10 equiv. of the oxidant were needed with this procedure whereas the previous the

TPAP/NMO combination could be applied in a ratio of 0.07:1.5 equiv. in relation to 1.0 equiv. of the alcohol. The TPAP loading could be as low as 1 mol% without affecting the *ees*, but reaction times were longer. Twice more the amount of chiral amine than that used previously was also required to obtain comparable yields and *ees*.

The mechanism proposed for these reactions is shown in Figure 7. The oxidation/asymmetric cyclopropanation cascade of allylic alcohols and bromomalonate shown above could also be achieved with a different oxidant combined with the same chiral amine (**5**), as reported by the Rueping group around the same time [37]. In this case manganese dioxide was used as a heterogeneous oxidant and triethylamine as base, to provide the required nucleophile activation. No oxidation of the amine was observed under these conditions and the oxidant was removed by filtration. However, 10 equiv. of the oxidant were needed with this procedure whereas the previous the TPAP/NMO combination could be applied in a ratio of 0.07:1.5 equiv. in relation to 1.0 equiv. of the alcohol. The TPAP loading could be as low as 1 mol% without affecting the *ees*, but reaction times were longer. Twice more the amount of chiral amine than that used previously was also required to obtain comparable yields and *ees*.

The TPA/NMO system combined with diphenylprolinol TMS-ether **5** was later selected by the Rueping group to obtain bi- and tricyclic chromenes from (*E*)-2-(2-nitrovinyl)phenols (**16**) and propargyl alcohols (**17**) [38]. Bicyclic chromenes (**18**) were initially obtained *via* a triple domino reaction consisting of oxidation/oxo-Michael/Michael addition (Figure 8). The presence of a basic by-product did not deteriorate the results and the chromenes were obtained in high yields and excellent *ees* (91-99%). They could subsequently be converted *via* a Michael/aldol condensation double domino reaction into the more complex tricyclic skeletons upon reaction with cinnamaldehydes in the presence of the same chiral amine. In this case addition of triethylamine was beneficial, and only one isomer of the products was obtained. Aryl-, cycloalkyl- and alkyl-substituted propargyl alcohols could be used as substrates for the triple domino reaction. The 4-(*H*)-chromene unit produced in this way is present in many biologically active natural products and medicinal compounds.

As a follow-up to these developments, the Rueping group reported an example of asymmetric conversion catalysis as a means to obtain dihydroquinolines from various combinations of an allylic alcohol with a 2-aminobenzyl alcohol, in 2014 [39]. Two oxidation reactions take place concurrently and independently in one pot and the aldehydes produced react further *via* an aza-Michael addition/aldol condensation/elimination reaction to yield the desired products in high yields and excellent *ees* (Figure 9). In this variation of synergistic catalysis, the same oxidant/catalyst combination reacts with both substrates, the allyl alcohol and the benzyl alcohol, albeit independently. This is followed by iminium-enamine catalysis and an elimination reaction. This complex cascade allows five different transformations to occur in one pot, without outside interference, in a highly efficient manner. The TPAP/NMO oxidative system proved again to be very versatile and chiral amine **5** was also found to be the best chiral catalyst out of a few tried. Molecular sieves, which are known to improve the efficiency of TPAP/NMO oxidations, were also used and probably promoted the elimination step as well.

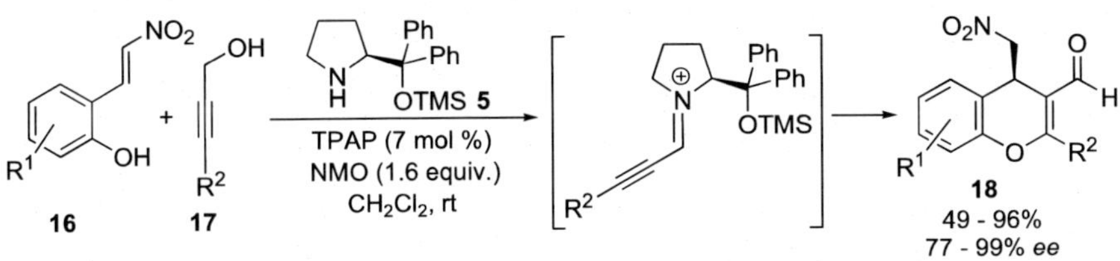

Figure 8. Oxidative cascade reactions for the synthesis of 4*H*-chromenes.

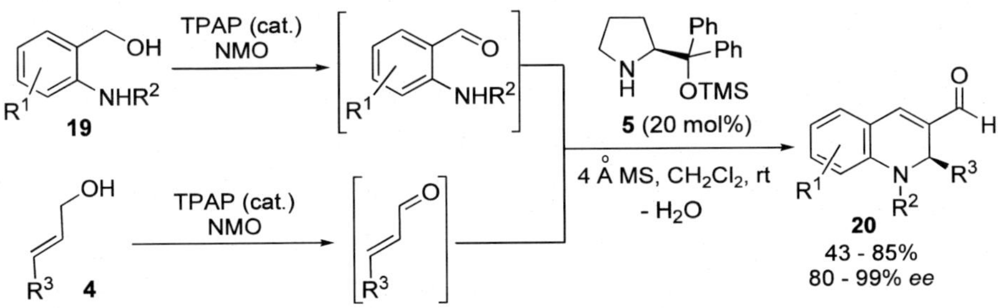

Figure 9. Synthesis of dihydroquinolines *via* asymmetric convergent catalysis.

In 2013 Kobayashi and co-workers reported novel examples of relay catalysis with heterogeneous catalysts capable of activating alcohols not only for an initial oxidation, but also for subsequent enantioselective reactions, in a single "one-pot" operation [40]. The catalysts had different active sites, which made bifunctional activation of reagents possible. Their activity was found to be strongly dependent on the way they were prepared, with the best ones being those in which nanoparticles (NPs) were layered over the organocatalyst. Previously the group had demonstrated that polymer incarcerated (PI) Au-Pd nanoparticles (PI(Au/Pd)NPs) having carbon black as secondary support, had high activity for aerobic oxidation of secondary allylic alcohols [41]. Similar reactivity was expected for primary alcohols. Out of a few bifunctional catalysts prepared, the best results were obtained with one prepared as follows: a Jϕrgensen-Hayashi type-organocatalyst (**21**) was suspension polymerized with two acrylamides, **22** and **23**, into polymer beads (Figure 10). The beads were then cross-linked with a styrene-based copolymer and treated with Au and Pd salts. The final result of this process was a polymer containing bimetallic Au-Pd alloy NPs, PI(Au/Pd)NP-CO catalyst **24**, with a gold content of 2 mol%. This catalyst promoted the aerobic oxidation/Michael addition of dibenzyl malonate to allylic alcohols in one pot, in high yields and *ees* (74-91%) (Figure 11).

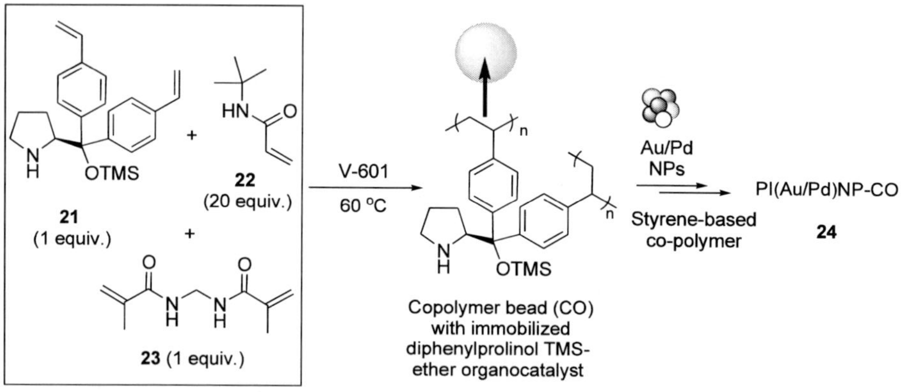

Figure 10. The synthesis of the PI(Au/Pd)NP-CO hybrid catalyst.

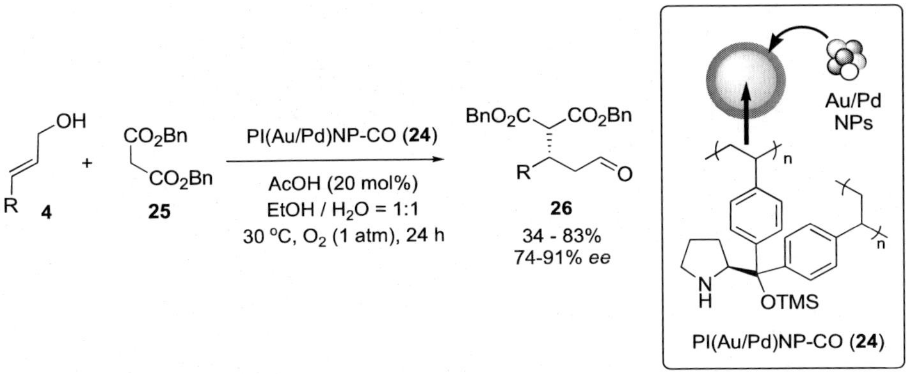

Figure 11. Relay catalysis by the PI(Au/Pd)NP-CO hybrid catalyst.

More recently Singh and co-workers used a cascade approach to obtain cyclohexanes bearing a variety of substituents in a total regio-, diastereo- and enantioselective fashion [42]. An oxidative Michael-aldol domino reaction (a triple domino) was possible when allylic alcohols **4** were combined with γ-nitroketones **27**, in the presence of base and TPAP/NMO, to oxidize the alcohol *in situ* to an α,β-unsaturated aldehyde which then underwent further reactions. The nature of the solvent selected was very important and the best results were obtained when the reactions were performed in a 2:8 mixture of 1,2-dichloroethane (DCE) and α,α,α-trifluorotoluene. As chiral catalyst diphenylprolinol TMS-ether **5** proved to be the best and 1,4-diazabicyclo[2.2.2]octane (DABCO) worked well as co-catalyst. The yields of cyclohexanes **28** obtained in these triple domino reactions were slightly lower than those obtained when the reaction was performed starting from preformed aldehydes, but the *ees* were similar. This method, which involves relay oxidation (OX), iminium catalysis (Im) and base catalysis (B), did not suffer from the drawbacks associated with some iminium–enamine cascades involving intramolecular aldol reactions, in which there is also dehydration with concomitant loss of two stereogenic centres, an unwanted feature in this case (Figure 12). The method could be applied to the formal enantioselective synthesis of ABT-341, a highly selective and orally bioavailable dipeptidyl peptidase IV (DPP4) inhibitor. DPP4 inhibition is useful for the therapy of type II diabetes.

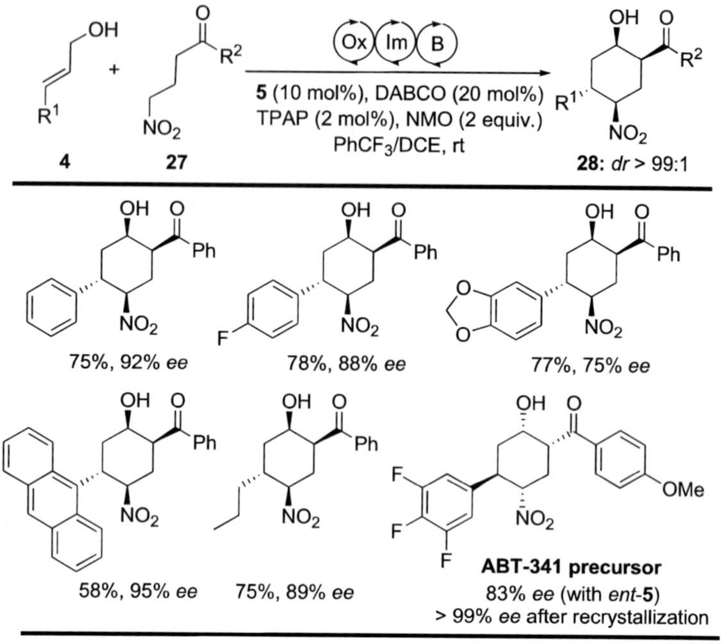

Figure 12. Oxidative Michael-aldol triple domino reaction for the enantioselective synthesis of highly substituted cycloalkanols.

2.2. Methods Based on Hydrogen Borrowing

In hydrogen borrowing, a hydrogen autotransfer process, the alcohol is temporarily converted into a carbonyl compound by metal-catalysed removal of hydrogen. The carbonyl compound undergoes further reactions *in situ* and at the end the metal returns the hydrogen to the newly produced compound. The final compound may also be an alcohol, but structurally more complex than the starting material, or a different substance. There is no change in the oxidation state of one of the starting materials. This process bypasses the need to have an oxidation reaction, followed by further synthetic steps and finally a reduction. It bypasses the need for stoichiometric oxidants, since the metal and other catalysts are used in catalytic amounts, hence there is waste reduction.

Although hydrogen borrowing has been known for some years, for example for alcohol amination [43], there have been very few applications in enantioselective synthesis [27, 44, 45]. In an early example, Williams and co-workers coupled benzyl alcohol with a stabilized phosphonium ylide, $Ph_3P=C(Me)CO_2Et$, using a borrowing hydrogen strategy [46]. This was achieved with $[Ir(cod)Cl]_2$ and (*S*)-BINAP as chiral ligand. Although one single example was described in this paper, after a screen for a suitable metal catalyst and the conversion to 2-methyl-3-phenyl-propionic acid ethyl ester was only 68%, the product could be obtained in 58% yield and 87% *ee*. According to recent reviews, only in 2013 appeared the next reports of synthetic methods in which high *ees* were obtained [44, 45]. This breakthrough was made when the first multicatalytic enantioselective approach was reported by Quintard, Rodriguez and co-workers [47]. The success of this approach, in which allylic alcohols (**4**) were reacted with β-ketoesters (**29**), relied not only on the fact that the metal complex used was able to selectively dehydrogenate the alcohol and to transfer the hydrogen back to the aldehyde after its reaction(s) with the second substrate, but also on the fact that the metal complex was compatible with chiral amines, so that iminium-enamine chemistry could be used. In addition, the reaction temperature was important. It was desired to keep it below 40 °C, to favour enantioselective processes. The metals most used for hydrogen-borrowing reactions until then were Ru, Ir and Rh, operating at high temperatures, unfavourable for enantioselective reactions [48]. An iron cyclopentadienone complex was however found, the Knölker complex (- **30**) [49], known to be useful for hydrogen transfer reactions, which could be adapted for a borrowing hydrogen-based multiple enantioselective catalysis strategy. In this process, Me_3NO, employed in catalytic amounts, was selected to activate the iron complex by CO de-coordination, giving rise to the active dehydrogenation catalyst **31**. Chiral amine - **5** was found to be compatible with the iron catalyst, and thus an enantioselective domino oxidation/ Michael addition/reduction cascade was possible, in which highly functionalized alcohols were obtained, *via* hydrogen borrowing. A plausible catalytic cycle for the relay process is shown in Figure 13. The products of the reaction exist as an equilibrium mixture of free alcohol (- **32**) and hemiacetal (- **33**) (Figure 14). Another interesting feature of the reaction is that the effect of the catalysts was truly synergistic, since if one of the components, the iron catalyst, Me_3NO, or the chiral amine, was removed from the mixture, no reaction was observed at all.

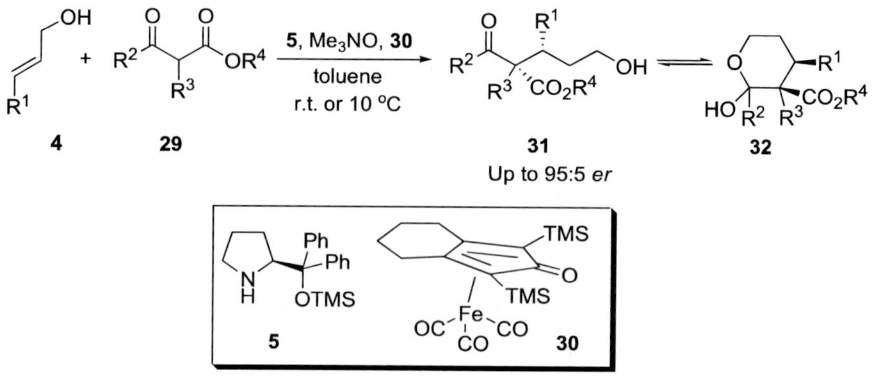

Figure 13. Functionalization of allylic alcohols by dual iron-prolinol catalysis.

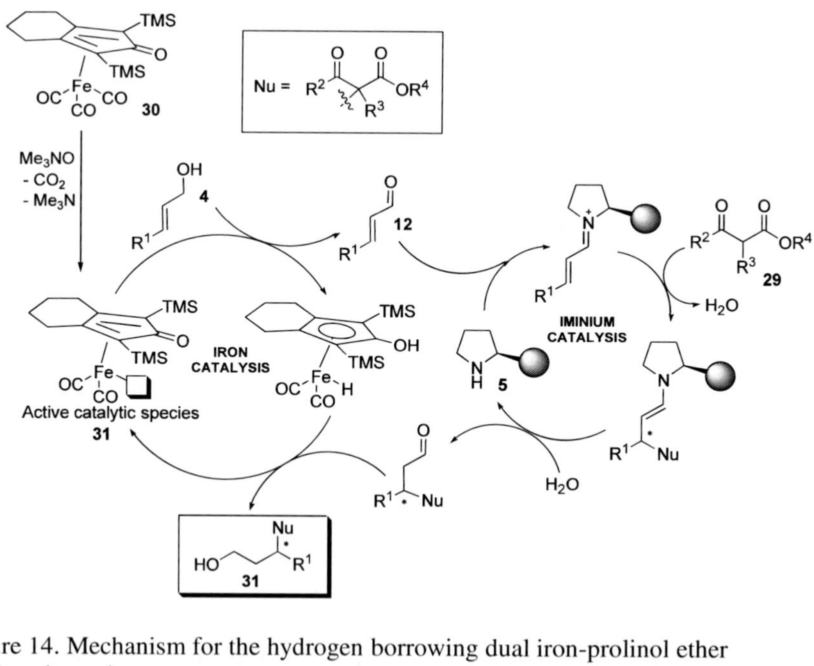

Figure 14. Mechanism for the hydrogen borrowing dual iron-prolinol ether catalyzed reaction.

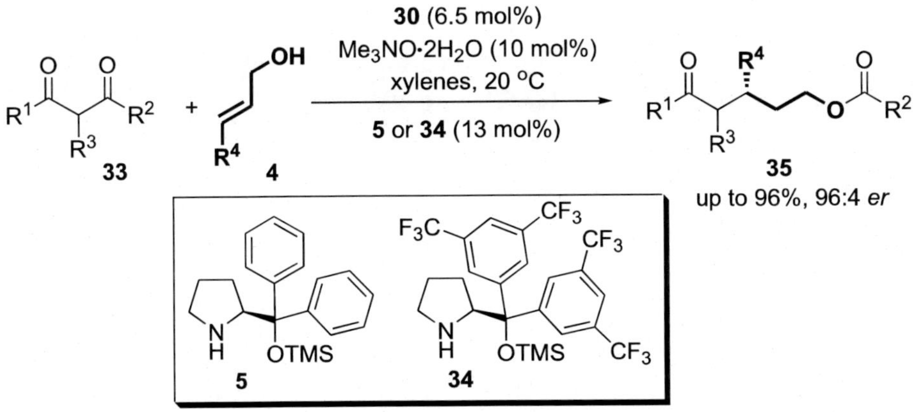

Figure 15. Quadrupole cascade in the synthesis of functionalized γ-chiral alcohols.

An extension of this hydrogen borrowing strategy to reactions of β-diketones, made possible the synthesis of functionalized γ-chiral alcohols in up to 96% yield and 96:4 *er* [50]. Optically active functionalized 3-alkyl pentanol fragments are important structural units of many biologically active natural products and even of fragrances, e.g., Doremox, a constituent of the rose scent which is commercialized by Firmenich. The methods available to produce these units rely on multistep sequences and the control of the absolute configuration at the γ-position in a γ-chiral alcohol is a challenge. The methodology developed by Quintard, Rodriguez and co-workers involves a dual borrowing hydrogen/iminium catalysis/retro-Claisen/fragmentation quadruple cascade reaction, which is in fact a formal chemoselective reductive insertion of a 3-substituted allylic alcohol into a C(O)–C bond of the 1,3-diketone substrate. The reaction was possible with catalysis by the same iron complex utilized for the reactions of β-ketoesters described above, although for the organocatalyst there were two alternatives possible, i.e., prolinol amines **5** or **33** (Figure 15).

The same reaction was tried on substrates in which one of the keto functions of the β-diketone was replaced by a stronger electron-withdrawing group, e.g., by a nitroketone or by a trifluoromethylketone, but in these cases no reaction was observed. The authors attributed this behaviour to a possible coordination of these groups to the iron atom of the catalyst, thus preventing the initial hydrogen transfer step from taking place. The method developed was applied to a short synthesis of the most odorant isomer of Doremox, the (2*S*,4*R*) isomer (Figure 16). Previously described enantioselective methods involved six- to nine-step synthetic routes [51]. The new methodology involving hydrogen borrowing is an atom-, step- and redox-economical way of producing functionalized γ-chiral alcohols.

Zhao and co-workers were also interested in applications of the hydrogen borrowing methodology. They described the first example of catalytic asymmetric amination of alcohols *via* borrowing hydrogen in 2014 [52]. This was possible using cooperative catalysis between a chiral iridium complex (- **39**) and a chiral phosphoric acid, 1,1'-bi-2-naphthol-derived **40**, commonly known as TRIP (Figure 17). A wide range of secondary alcohols with alkyl-alkyl or alkyl-aryl substitution were compatible with this process, as well as various primary amines. Enantioenriched secondary amines could be obtained in 64-98% yields and 69-97% *ees*, despite the fact that a high refluxing temperature was needed. The *ees* could be raised by lowering the temperature, but at lower temperatures the yield dropped. Products bearing electron-donating substituents were obtained with higher *ee* values, which suggests that the electronic properties of the substrates played a role in the chirality transfer process. One limitation was that only aromatic amines worked well in this reaction. The authors proposed a mechanism involving the formation of an iridium hydride species, which could be observed by ^1H NMR spectroscopy. Interestingly, when the reaction was performed in the absence of the chiral phosphoric acid the Ir-H signal was not observed, which supports the fact that the chiral phosphoric acid plays a crucial role in the mechanism. In fact, when the reaction was attempted on a model substrate in the absence of this acid with all other conditions remaining the same, the yield of product was only 13%. Presumably TRIP plays a role in the imine condensation as well as in activating the imine intermediate as an iminium for attack by the chiral iridium hydride complex.

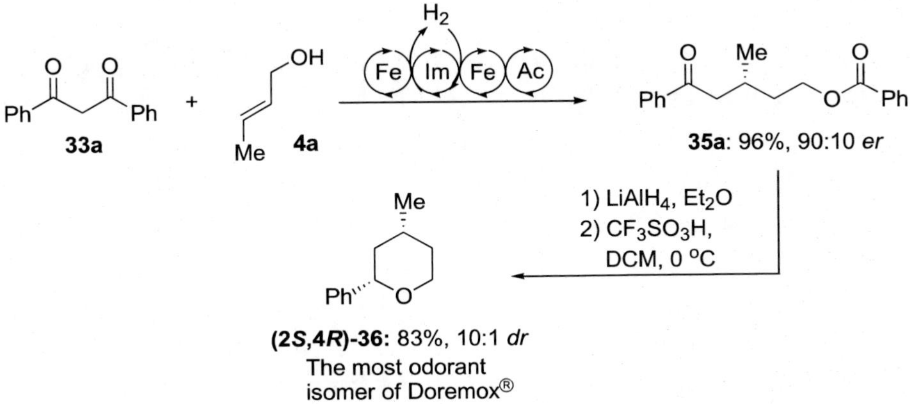

Figure 16. Synthesis of Doremox *via* hydrogen borrowing.

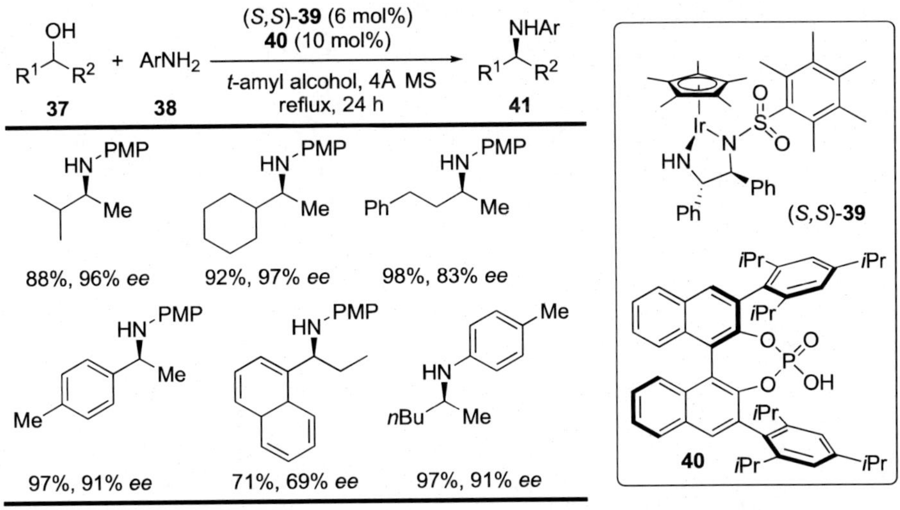

Figure 17. The first catalytic asymmetric amination of alcohols *via* hydrogen borrowing.

In this reaction chiral induction was also dependent on the chirality of the phosphoric acid selected, since when simple racemic diphenyl phosphoric acid was applied as catalyst in combination with the tosyl analogue of (*S,S*)-**39**, the *ee* obtained was lower than 10%. The same tosyl-substituted iridium complex employed in combination with TRIP, afforded an *ee* of 81% in the same model reaction.

The Zhao group was also the first to describe the dynamic kinetic resolution (DKR) of alcohols *via* an asymmetric amination reaction using hydrogen borrowing [53]. DKR is a particularly useful strategy for synthesis, since the two enantiomers of a racemic mixture can be converted into an enantioenriched product, which is not possible using traditional kinetic resolution, in which only one of them is converted and the maximum theoretical yield is limited to 50%. To achieve DKR it is necessary that there is rapid racemization of the substrate (or of an intermediate), prior to the enantio-determining step. Alternatively, the racemic substrate has to be converted into an achiral intermediate just before the enantio-determining step. This approach could be used for synthesis of chiral amines (Figure 18). α-Branched secondary racemic alcohols were converted into amines bearing two adjacent chiral centres in a highly enantio- and diastereoselective fashion through the cooperative action of catalytic amounts of iridium catalyst (*S,S*)-**39** and chiral phosphoric acid **40**. Surprisingly, very high asymmetric induction could be obtained despite the very high reaction temperature required.

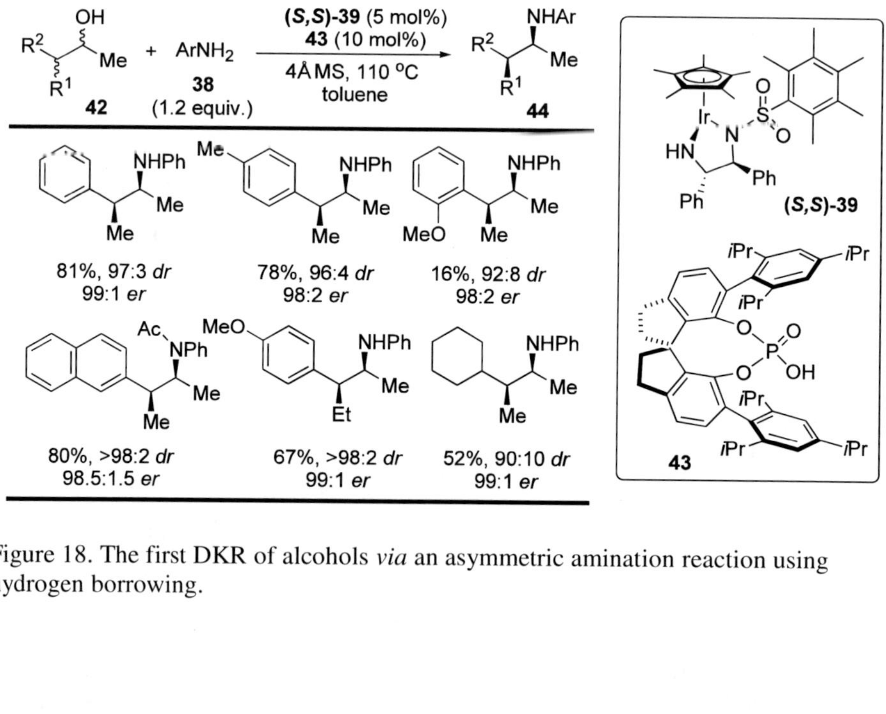

Figure 18. The first DKR of alcohols *via* an asymmetric amination reaction using hydrogen borrowing.

In this highly stereoconvergent process, the racemic alcohol present as a mixture of four stereoisomers could be transformed into a highly enantioenriched chiral amine as a result of a DKR (Figure 19). The alcohol has two stereocentres which can be racemized first *via* an oxidation to a ketone and secondly *via* tautomerization of the iminium intermediate **46** and **48** through an enamine **47**. The chiral catalyst had to differentiate not only between **46** and **48**, but also between two enantiotopic faces of the imine moiety so that the highly diastereo- and enantioselective synthesis could be achieved.

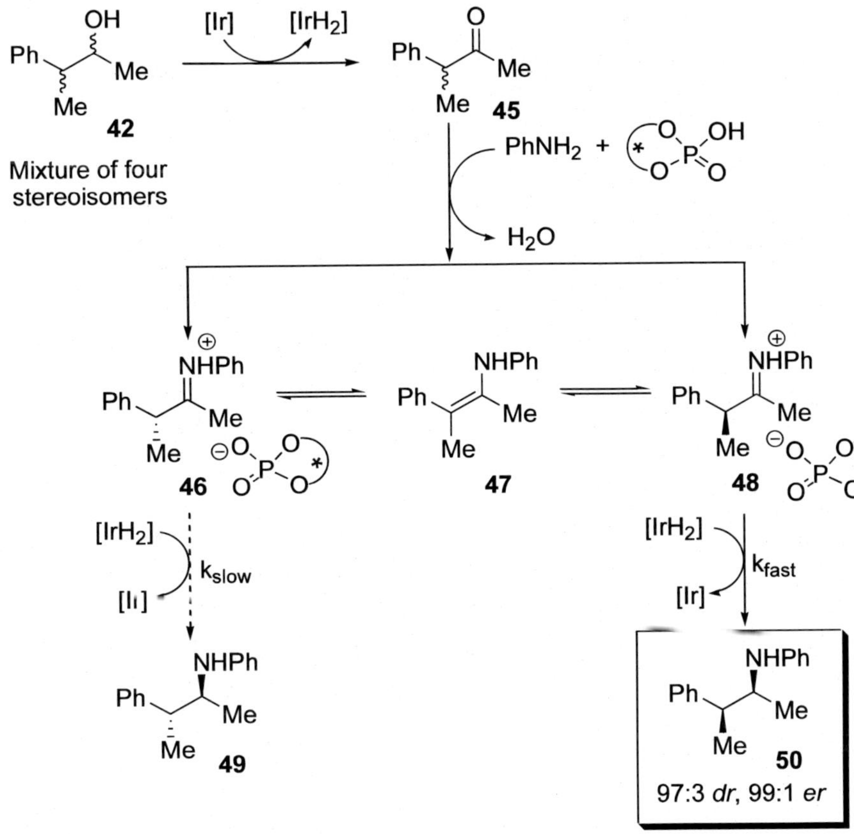

Figure 19. Mechanism proposed for the DKR of alcohols.

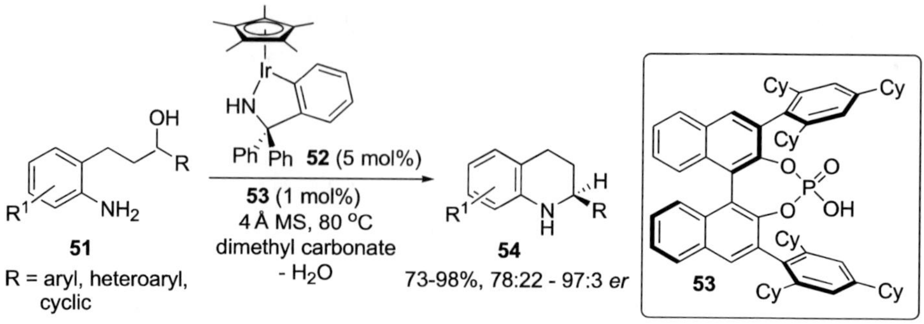

Figure 20. Synthesis of tetrahydroquinolines *via* hydrogen borrowing.

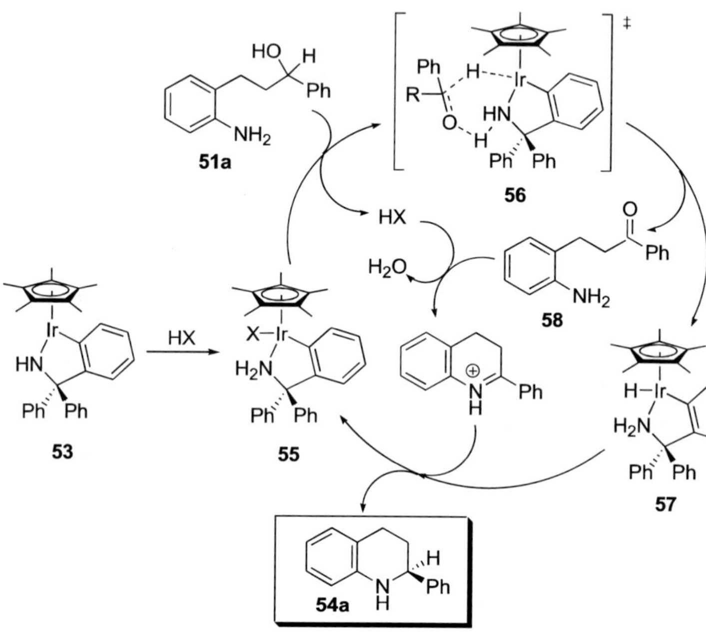

Figure 21. The mechanism proposed for the synthesis of tetrahydroquinoline **54a**.

Hydrogen borrowing has also been used for a highly sustainable synthesis of tetrahydroquinolines by Zhao and co-workers [59]. Starting from an amino alcohol with an appropriate length of tether, these substances could be obtained *via* an oxidation/intramolecular amination cascade with the cooperative catalysis of an achiral iridacycle complex (**52**) and a BINOL-derived chiral phosphoric acid bearing bulky 2,4,6-tricyclohexylphenyl

(2,4,6-$Cy_3C_6H_2$) substituents at the 3,3´-positions (**53**). The reactions afforded products in very high yields and *ees* (Figure 20). The hydrogen borrowing method is much more sustainable than other known procedures to obtain these substances, since it requires no external oxidants or reductants and the only by-product is water. In contrast, asymmetric hydrogenation requires, besides the catalysts, H_2 gas at high pressures, asymmetric hydrogen transfer requires an Hantzsch ester or a benzothiazolidine as reductant and dearomative arylation has a boronic acid and salts as by-products. A mechanism for this redox-neutral amination of alcohols is shown in Figure 21.

2.3. Sequential "One-Pot" Reactions

This section covers reactions in which there is stepwise addition of reagents or catalysts, or in which there is a change in reaction conditions, before the products are isolated. The first example of an enantioselective mixed organocatalyst-metal catalysed sequential cascade was reported by Kudo and co-workers [55]. It was developed soon after the group showed that enantiomerically enriched α-oxyaminated aldehydes could be obtained directly from alcohols when TEMPO and molecular oxygen were employed for alcohol oxidation in the presence of a resin-supported peptide catalyst (Figure 5, section 2.1) [34]. In order to achieve other transformations of alcohols following a similar approach, but avoiding the concurrent formation of products resulting from the reaction of TEMPO with the aldehyde, the authors decided to immobilize TEMPO on a separate resin bead. The two catalysts were used jointly for a direct cross-aldol reaction between *in situ* prepared aldehydes with ketones. The basis for this approach were previous studies by the group, which showed that what would otherwise be incompatible acidic and basic catalysts could be applied simultaneously in a single pot reaction, provided that they were pre-immobilized on separate resin beads [56]. Thus, for this sequential cascade, a peptide catalyst was attached to a polyethylene glycol-polystyrene (PEG-PS) resin and TEMPO was immobilized on an amino terminated PEG-PS resin *via* a 4-carboxy group using conventional condensation chemistry. The alcohol oxidation proceeded smoothly in the presence of air when these two catalysts were mixed with copper(I) chloride and 2,2´-bipyridine. The authors also found that if glycine residues were introduced between TEMPO and the PEG-PS resin, they could be used to complex the copper salt while bringing it into close proximity to the TEMPO moiety. The conversion obtained in this case was a function of the number of glycine units used, but it remained low. The reaction was performed in a Tris-HCl (pH 7.4) buffer, to suppress the non-enantioselective background reaction caused by 2,2´-bipyridine acting as a base. The best results were obtained when, during preparation of the catalysts, the resins were pre-treated with a DMF solution containing excess copper salt and 2,2´-bipyridine. During this process copper became adsorbed and the new catalysts oxidized alcohols at much higher rates. The best option was catalyst **61**. It contained a triglycyl tether and it could adsorb 1.0 mol equiv. of copper complex per TEMPO, as determined by gravimetric analysis. In contrast, a simple TEMPO-bound PEG-PS resin could entrap only 0.3 mol equiv. of the complex. Resin-supported peptide D-Pro-Tyr(*t*-Bu)-Phe (**62**) was found to be the best catalyst for the aldol reaction (Figure 22). It could be recovered by filtration and re-used up to 8 cycles without any loss in performance.

Jang and co-workers studied the addition of malonates to allylic alcohols using TEMPO/molecular oxygen for the alcohol oxidation [57]. This reaction had been previously studied by Rueping and co-workers (Figure 6), who, using the dual catalytic system MnO_2/TPAP/chiral pyrrolidine **5**, achieved the addition under domino conditions. In order to use the new oxidant, Jang and co-workers found that a sequential reaction was necessary. Copper(I) chloride, employed as catalyst, was added to the reaction vessel at the start with TEMPO. The oxidation was complete within 1 h, after which malonate and chiral pyrrolidine **5** were added to the reaction vessel. The Michael adducts were obtained in good to high yields (43-85%) and very high *ees* (80-96% *ee*) (Figure 23). The addition of benzoic acid improved considerably both the yields and *ees* of this sequential oxidation/Michael addition cascade and in the presence of oxygen only 10 mol% of the copper catalyst and TEMPO were needed to obtain these results. The authors also observed that small amounts of α-oxyaminated Michael adduct were produced (<5%), as found previously by Kudo and co-workers during the oxidation of saturated alcohols (Figure 5, section 2.1). These findings prompted further studies and a "one-pot" method for the sequential oxidation/Michael addition/α-oxyamination of allylic alcohols was developed. The oxidation step was performed as in the 2-step version, and once complete, additional TEMPO (2 equiv.) was added to the reaction vessel together with the malonate and the acid. For this triple cascade, adamantane-1-carboxylic acid replaced benzoic acid, since it was found to be a better additive. The products were obtained in moderate to good yields and very high stereoselectivities.

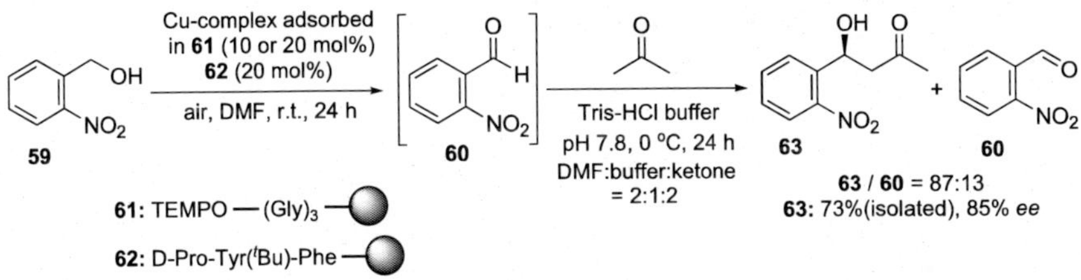

Figure 22. Sequential one-pot reaction catalyzed by by resin-immobilized TEMPO and a resin-immobilized peptide.

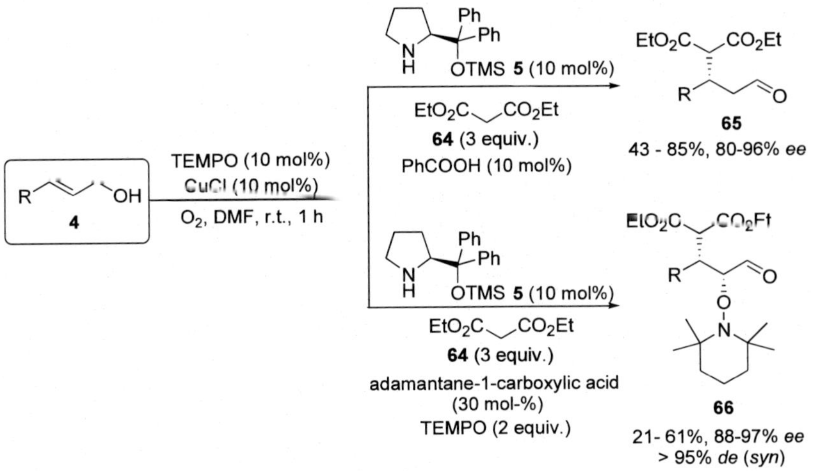

Figure 23. Sequential oxidation–Michael addition cascade.

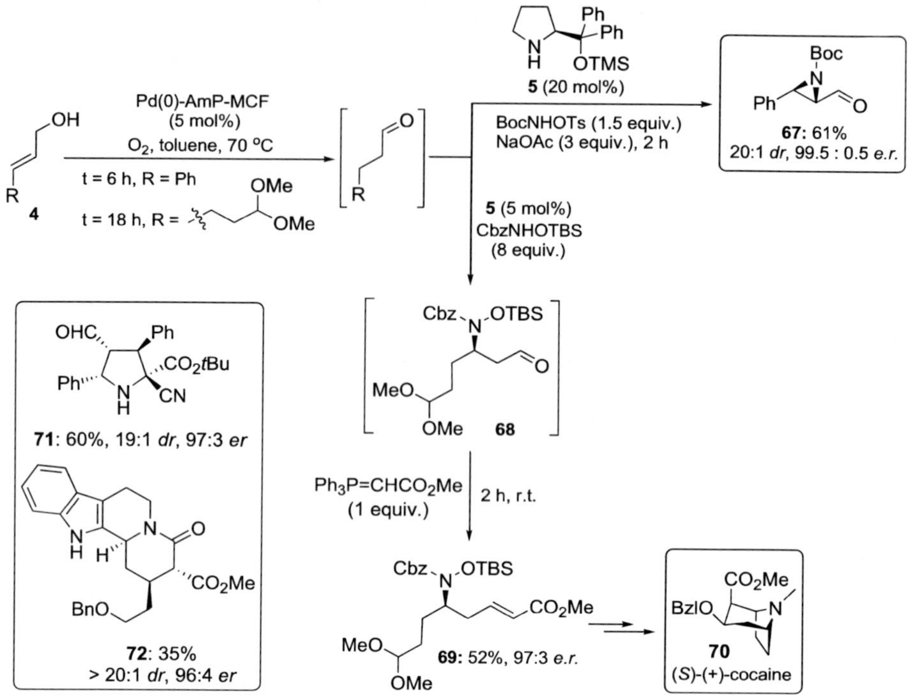

Figure 24. Examples of Pd-chiral pyrrolidine ether cooperative catalysis.

In 2014, Johnston, Córdova and co-workers described dual palladium-organocatalysis as a means to obtain enantiopure molecules with a variety of functionalities [58-60]. Sequential cascade reactions with different levels of complexity were reported. A palladium catalyst, previously used for aerobic alcohol oxidations, was used. It contained Pd supported on an aminopropyl-functionalized mesocellular foam (Pd(0)-AmP-MCF) [61, 62]. When combined with chiral pyrrolidine **5**, one-pot cascade sequences were developed to obtain not only chiral aziridine **67**, but also Michael adduct **68**, which could be trapped *in situ* and used as intermediate for further reactions, including a "one-pot" synthesis of (*S*)-cocaine (**70**) (Figure 24). This chemistry could also be applied to the synthesis of highly substituted chiral pyrrolidines, e.g., **71** and a tetracyclic intermediate (**72**) for the total synthesis of secologanine tryptamine alkaloids.

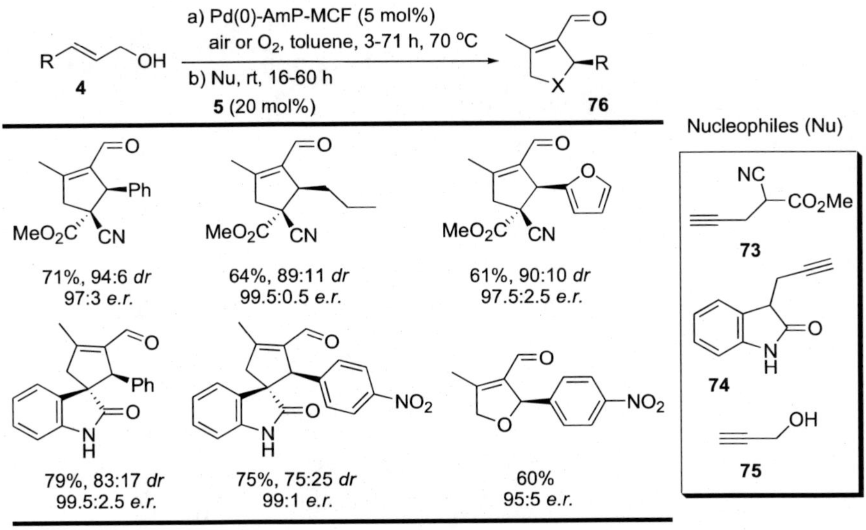

Figure 25. Asymmetric synthesis of polysubstituted carbo- and heterocycles.

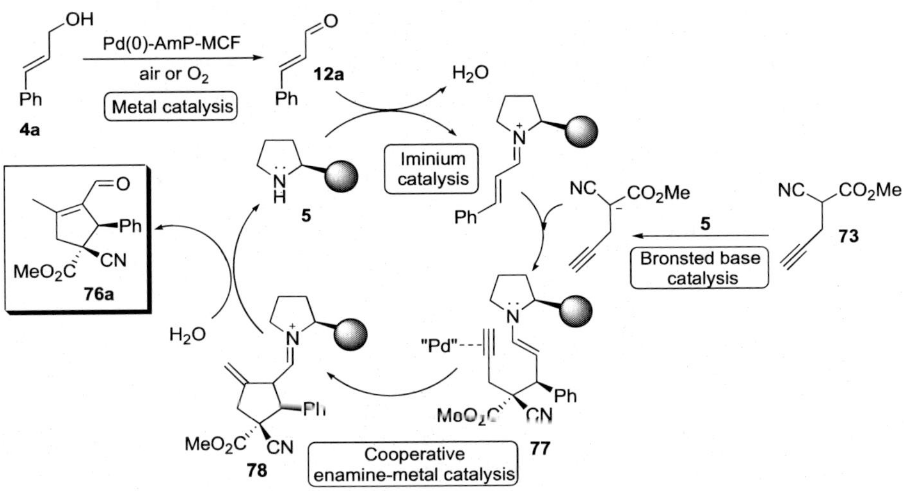

Figure 26. Mechanism proposed for the dual Pd-chiral pyrrolidine ether-catalysed synthesis of carbocyles.

The Córdova group reported other interesting examples of cooperative catalysis. Cyclopentenes and spirocycle oxindoles bearing a quaternary stereocenter, as well as chiral dihydrofurans, could be made using similar strategies (Figure 25) [62]. For this, allylic alcohols were reacted either with propargylic cyanoacetate **73**, propargyl isoindolinone **74**, or propargyl alcohol **75**. In reactions in which **74** and **75** were the nucleophiles, benzoic acid was used as additive. A probable mechanism is illustrated in Figure 26. For example, in the synthesis of cyclopentenes, the heterogeneous palladium catalyst promotes the aerobic oxidation of the allylic alcohol, after which there is dual Pd/**5**-co-catalyzed carbocyclization between the *in situ* generated enal, reacting as an imine and the propargylic cyanoacetate, which behaves simultaneously as a nucleophile after deprotonation by the amine and as an electrophile through its triple bond which is more electron deficient as a result of coordination to the palladium catalyst (Lewis acid activation). When the nucleophile used was propargyl isoindolinone **74**, spirocycle indoles could be obtained (Figure 26).

To the process shown in Figure 24, it was also possible to add another synthetic step in a sequential manner, prior to the isolation of the products, and to produce chiral polysubstituted aziridines with high stereoselectivities. This was a novel aerobic oxidation/Michael/carbocyclization/aziridination sequence. The heterogeneous palladium-chiral amine catalyst combination employed in these reactions proved to be very versatile. It could be applied to catalyse both oxidation and reduction steps in the same one-pot cascade sequence in the preparation of polysubstituted cyclopentanes **81** (Figure 27).

To achieve this, once the initial oxidation/Michael/carbocyclization cascade sequence was complete, the reaction vessel was evacuated and the oxidizing atmosphere was replaced by hydrogen gas. The reduction of the double bond of the cyclopentene was 1,4-regiospecific and highly enantioselective, and only one product was obtained, containing a quaternary all-carbon stereocenter, in greater than 99.5:0.5 *er*. The aldehyde and the ester functions in **79** could also be reduced *in situ* subsequently, either jointly with sodium borohydride or, when $NaCNBH_3$ was used as the reducing agent, to the alcohol ester. This was the first example of amine-catalyzed stereoselective hydrogenations of enals in which hydrogen gas was used instead of a Hantzsch ester. The species hydrogenated was presumably the conjugated pyrrolidinium cation. The importance of the presence of the chiral amine and the fact that synergistic catalysis was operating could be inferred from the fact that when the palladium-catalyzed hydrogenation of, **76b**, i.e., **76,** X= C(CN)(CO_2Me),R = 4-MeOC_6H_4, was attempted in the absence of the chiral amine, only trace amounts of product were obtained after 72 h.

Mechanistic studies were later performed by means of density functional theory calculations on the carbocyclization step [63]. In combination with other experiments, they suggested that the C−C bond-forming reaction transformation is probably a two-step carbocyclization of enaminynes catalyzed by a Pd(II) species, i.e., initially the nucleophilic enamine attacks the electrophilic Pd(II), a step which is followed by an insertion into the alkyne bond to afford the carbopalladation intermediate. The alcohol to cyclopentene conversion and related reactions, are examples of "Pd oxidase catalysis," i.e., a palladium-catalyzed selective direct aerobic oxidation of alcohols, which proceeds without the need for an active redox cocatalyst for dioxygen-coupled catalytic turnover [64]. This "metal/organocatalyst cooperative catalysis" may find several applications in sustainable chemistry.

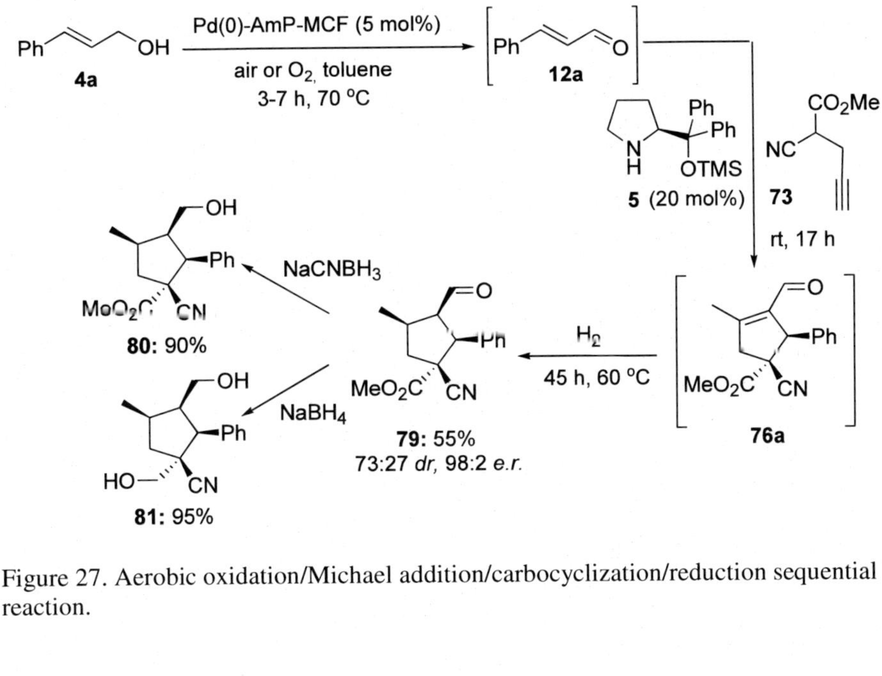

Figure 27. Aerobic oxidation/Michael addition/carbocyclization/reduction sequential reaction.

Kobayashi and co-workers developed bifunctional heterogeneous materials which could also be applied in sequential cascade reactions. The successes obtained with the polymer incarcerated Pd/Au NPs copolymerized with a chiral pyrrolidine (the PI(Au/Pd)NP-CO catalyst) (Section 2.1, Figures 10 and 11) [40] prompted further studies in this area. One shortcoming of this catalyst had been found to be the fact that it could not be recycled, since the chiral amine became deactivated under the aerobic conditions. Aiming to develop more robust catalyst which could be reused, the group developed novel ones in which the chiral amine was replaced by a chiral phosphoric acid, TRIP. Although TRIP activates substrates by mechanisms different from those of chiral amines, it is a catalyst which has been found to be useful for several reactions [65, 66], some of which have been described in previous sections of this chapter. The new hybrid catalysts were prepared in a manner similar to the previous ones, but the NPs were now Au-Pd alloys incarcerated in a polymer with carbon black (CB) as secondary support (PI/CB–Au/Pd) [40]. The required monomers and the catalyst precursor were subsequently copolymerized in the presence of these NPs, to give the final catalyst IOC–PI/CB(Au/Pd) (IOC = immobilized organocatalyst coated) (**82**) (Figure 28). Another cascade reaction initiated by alcohol oxidation (a TOP process) could be developed, combining the oxidation with a domino amination/intramolecular Friedel-Crafts reaction for the synthesis of chiral piperazines [67]. Oxygen was the terminal oxidant in this process too. The success of this cascade reaction was dependent on a few factors. Firstly, an entire domino process was not possible, because when the required *N*-aminoalkylpyrrole substrates were mixed initially with the alcohol, no reaction took place, presumably due to strong coordination of the primary amine function to the Au/Pd NPs deactivating the catalyst. The presence of water was also found to be required to promote the oxidation and $BnSCH_3$ was used as an additive, which was added once the alcohol oxidation was finished to deactivate the Au/Pd NPs. In its absence, the piperazine obtained was oxidized to its imine form. However, the new catalyst incorporating Au/Pd NPs could be recycled five times with less than 5% loss in yield or *ee* of the product obtained, and after six recycles the yield was still 85% and the *ee* 84%, in the synthesis of piperazine **84** (R = Ph).

Between cycles it was necessary to reactivate the catalyst, and this was done by treating it with an aqueous solution of H_2O_2 to remove $BnSCH_3$.

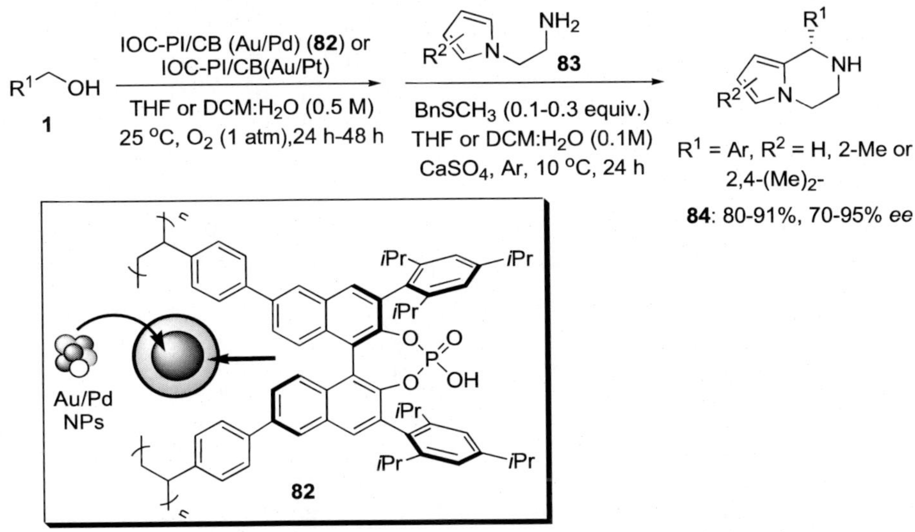

Figure 28. Synthesis of chiral piperazines *via* hybrid Au-Pd-CPA catalysis.

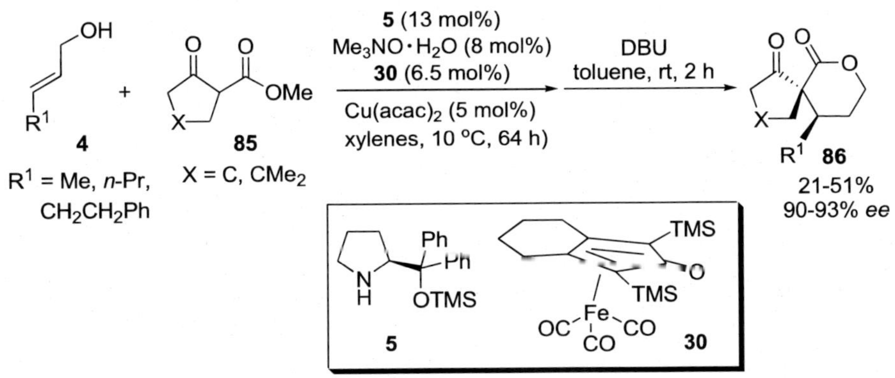

Figure 29. Synthesis of α-spirocycle δ-lactones *via* borrowing hydrogen.

Sequential processes were also a means that Quintard and Rodriguez found to synthesize α-spirocycle δ-lactones from β-keto esters and allylic alcohols using a borrowing hydrogen strategy [68]. These structural motifs are commonly encountered in many families of natural products [69]. This sequential cascade reaction was an extension of earlier work by the group (section 2.2), when it had been found that in the presence of the iron Knölker complex (**30**), a chiral pyrrolidine and catalytic amounts of Me_3NO, an enantioselective domino oxidation/Michael addition/reduction cascade was possible, leading to highly functionalized alcohols, which were obtained as equilibrium mixtures free alcohols and hemiacetals (Figure 29). Attempts made to obtain only ring closed compounds and even spirocycle products, revealed that the alcohols were rather sensitive and trying either acidic or mildly basic conditions led to the formation of decomposition products, presumably *via* side reactions of the unprotected ketones, when β-keto ethyl or *t*-butyl esters were used as starting materials. The β-keto methyl ester could however successfully converted into a spirolactone (- **86**) in the presence of base. Using a sequential approach, DBU was added once the domino reaction was complete, prior to any work-up and the lactones possessing a challenging quaternary stereocenter could be obtained in high *ees*. They were subsequently isolated as single diastereomers by chromatography (Figure 29). Adding $Cu(acac)_2$ to the system at the start resulted in a small improvement in *ee*, and the all spiro-δ-lactones were produced in ≥90% *ee* in 21 to 51% yield.

3. Cascade Processes Promoted by Metal Complexes and Metal Salts Only

Sigman and his group have been developing in recent years a different type of cascade reactions to transform alcohols into value-added products. They are redox-relay oxidative Heck reactions [70, 71]. They are innovative due to the fact that they allow the functionalization of allylic alcohols at a site remote to the functional group, a transformation which is normally difficult to achieve. These reactions are different from the classical Heck reactions and also from the typical oxidative Heck reactions. In the first, a Pd(0) catalyst is used to couple an halide or triflate to a compound containing an unsaturated C–C bond in the presence of base. In the second, a Pd(II) catalyst is employed and consequently, instead of the catalytic cycle starting with an oxidative addition of Pd(0) into a halide or triflate, it starts with a transmetallation between an organoboronic acid and the Pd(II) catalyst (Figure 30a). Oxidative Heck reactions usually do not require high temperatures or bases, but an oxidant, usually O_2, air, benzoquinone or $[Cu(OAc)_2]$, to re-oxidize Pd(0) to Pd(II) at the end of the catalytic cycle. In these two variations of the Heck reaction, after an initial migratory insertion step in which a C–C bond is formed, there is β-hydride elimination, and a Pd-H intermediate is produced, which reacts with the oxidant to regenerate the active Pd catalyst. If instead of undergoing β-hydride elimination, Pd remains coordinated to the alkene and the hydride reinserts at the opposite carbon, the palladium catalyst will be found at a position one carbon further down the alkyl chain (Figure 30b). This process is called "relay" or "chain walking" and produces an isomer of the original alkene. By successive (relay) migratory insertion and β-hydride elimination steps, the double bond migrates towards the hydroxyl group, which leads to an oxidation step, i.e., to **90**. In this redox economical process, the oxidation of the alcohol functionality to an aldehyde or ketone is the final step in the cascade, in contrast to the examples discussed so far. This mechanistic proposal was supported by mechanistic studies [72, 73].

In their first account, in 2012, Sigman and co-workers described the enantioselective conversion of allylic secondary alcohols into β-arylated ketones, and homoallylic secondary alcohols into γ-arylated ketones [70]. Although β-arylated ketones may be easily obtained by other methods, e.g., by conjugate addition to α,β-unsaturated ketones, but the enantioselective synthesis of γ-arylated ketones is far more difficult to achieve. A redox-relay oxidative Heck reaction proved to be an easy route. In their first developments of this methodology, aryldiazonium salts **91** were used as the arene source, as in the typical Heck–Matsuda coupling [71]. As ligand for palladium, chiral pyridine-oxazoline **92** proved to be very efficient, and the reactions proceeded in a highly enantioselective manner (Figure 31). The outcome was very sensitive to the double bond geometry of the starting alkene, and by using either one or the other geometrical isomer, enantiomeric products with opposite configurations could be obtained.

The reactions were also highly regioselective. When homoallylic alcohols were used as substrates, a small amount of β-arylated ketone (<5%), resulting from insertion at the alternate alkenyl carbon, was also observed. Primary alcohols yielded aldehydes in very high yields and *ees*. Substrates in which the double bond was an additional carbon away from the hydroxyl group yielded δ-substituted carbonyl products preferentially by relay (Figure 31). However, the selectivity was approx. 20% lower in these cases, and some γ-arylated ketone was also obtained.

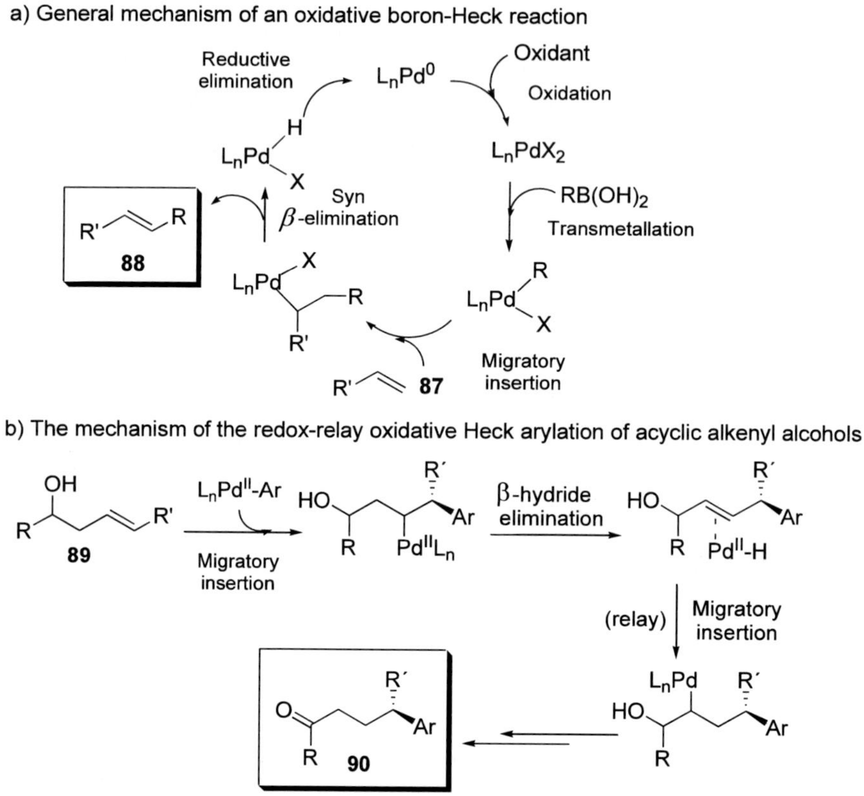

Figure 30. Comparison of the mechanisms of the oxidative Heck reaction (a) and a redox-relay oxidative Heck reaction (b).

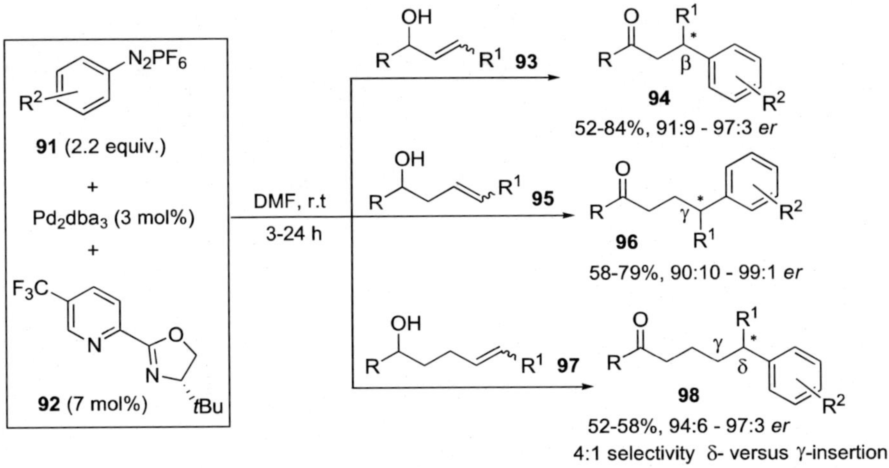

Figure 31. Examples of oxidative redox-relay Heck reactions.

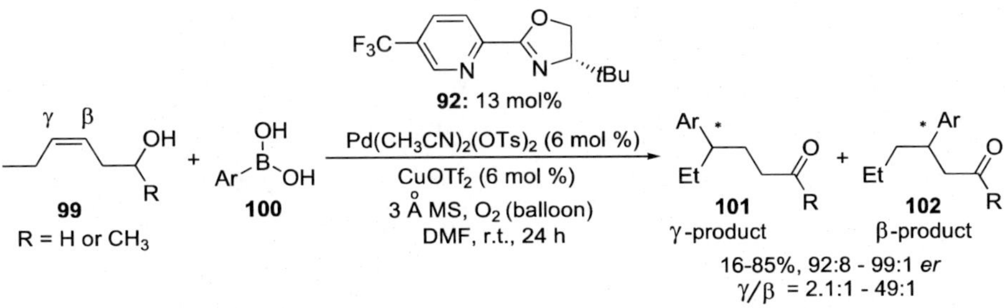

Figure 32. Oxidative redox-relay reactions with aryl boronic acids.

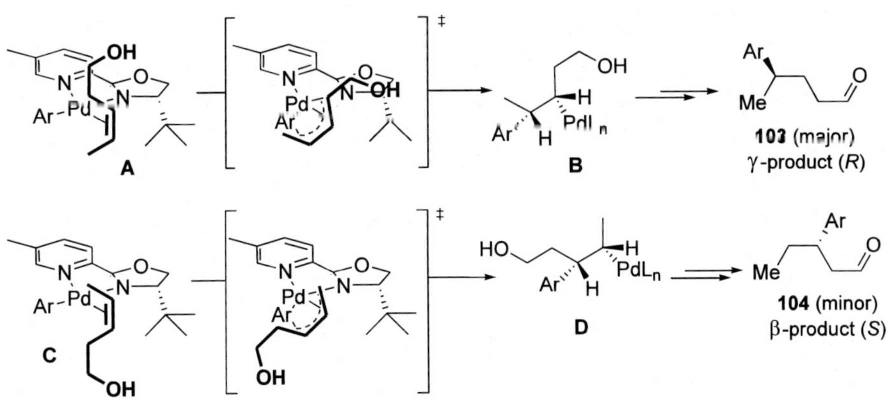

Figure 33. The mechanism proposed for the oxidative redox-relay Heck arylation of homoallylic alcohols.

Despite the very good results obtained, the use of aryl diazonium salts as coupling partners limits somehow the range of potential applications of this methodology. In order to expand the scope of their oxidative redox-relay Heck arylation, Sigman and co-workers changed the aryl coupling partners to aryl boronic acid derivatives (**100**) [74]. A cationic palladium salt was selected in this case as catalyst, and an atmosphere of O_2. Chiral pyridine oxazoline ligand **92** was also used to render the process enantioselective. In addition, copper triflate was also required as a co-catalyst, its presence caused a significant increase in yield. The function of O_2 or Cu(II) is to reoxidize the catalyst and no reaction takes place if palladium is removed from the system. Conversions also improved significantly when molecular sieves were added. In the past, poor conversions in palladium-catalyzed aerobic oxidations were associated with catalyst deactivation through Pd metal aggregation as a result of slow oxidation of Pd(0) to Pd(II). Using the optimized conditions, a variety of aryl boronic acids was reacted with homoallylic primary and racemic secondary alcohols affording high yields of arylated aldehydes or ketones respectively (**101/102**), in moderate to good yields and very high enantiomeric ratios (*ers*) (Figure 32). The site selectivity (γ *versus* β) for alkene functionalization varied from good to excellent. It was lower when the alkene was less electronically biased, suggesting that there is a remote electronic effect [70, 74]. Although in many enantioselective reactions the degree of chiral induction obtained is often very sensitive to the nature of the substrates, in these reactions this was not very much the case. However, the site selectivity for alkene insertion was found to depend both on the nature of the arylboronic acid and its substitution and also on the chain length of the alkenyl alcohol.

A Hammett plot obtained using Hammett σ-values of both *meta*- and *para*-substituted boronic acids *versus* the log of the ratio of site selectivity, showed a ρ-value of 1.42. The log of the ratio of site selectivity is proportional to relative rates of isomer formation. The ρ-value obtained suggests that the orientation of insertion is very sensitive to the electronic nature of the Pd–Ar species formed. The site selectivity was lower when more electron-rich aryl groups were used, i.e., when the Pd-aryl species became less electrophilic.

An interesting result obtained was the fact that the insertion reaction was highly enantioselective irrespective of which face of the alkene reacted with the catalyst, but it took place at different carbon atoms, i.e., the product of γ-insertion was *R*, e.g., **103,** whereas the product of β-insertion was *S,* e.g., **104**. The structure of the starting alcohols was also varied and it was observed that as the hydroxyl group became further removed from the double bond, site selectivity decreased, but the *ee* remained high, which suggests that the hydroxyl group is not involved in face selection. It was concluded that the enantioselectivity is controlled by steric repulsions between the oxazoline substituent and the alcohol-bearing alkene substrate. Plausible transition states, illustrated in Figure 33, show different orientations for **A** and **C** which take into account steric repulsions [72, 74]. The high level of enantioselection also supports the idea that the Pd–H species does not dissociate from the alkene, which is probably due to the electrophilic nature of the catalyst.

Mechanistic studies revealed that the C-O dipole of the alcohol moiety stabilizes the transition state [72]. The relay efficiency is due to an unusually smooth potential energy surface without high barriers, where the hydroxyalkyl-palladium species acts as a thermodynamic sink. This is the main factor which drives the reaction towards the formation of the final carbonyl compound. The mechanism of the redox-relay oxidative Heck reaction has also been studied by other groups. Wang and co-workers concluded that the palladium complex moves down the carbon chain by repeated complexation, and insertion/elimination reactions. At the end of this process, the very stable carbonyl product is obtained by enol-keto tautomerization [75]. Subsequently, Sigman and co-workers performed deuterium labelling experiments, which showed that although both (*E*)- and (*Z*)-isomers are formed, the alkene face selection remains constant. This supports the idea that palladium does not dissociate from the substrate during migration, but remains bound to one face only throughout the relay. Deuterium labelling also helped to elucidate the mechanism of carbonyl formation. If a standard Heck reaction was involved, an alcohol would be the final product. They concluded that tautomerization would be the last step if, at the end of the palladium insertion-elimination relay, a palladium-enol intermediate **A** was reached (Figure 34), containing a palladium-hydride bond, which could undergo oxidative deprotonation by DMF to yield **C**, and thus closing the catalytic cycle and releasing the product [73]. However, if instead, at the end of the palladium insertion-elimination relay, palladium hydride reinsertion occurs, **D** is obtained, and oxidative deprotonation by DMF releases product **E**. The formation of the palladium alkyl intermediate **D** was confirmed by labelling experiments. DFT calculations of the relevant transition structures at the M06/LAN2DZ+f/6-31+G* level of theory showed that the second pathway is favored by 5.8 kcal/mol [73]. These calculations suggested that there is generation of a low energy palladium–alkyl intermediate prior to product formation, and hence a tautomerization mechanism was discarded.

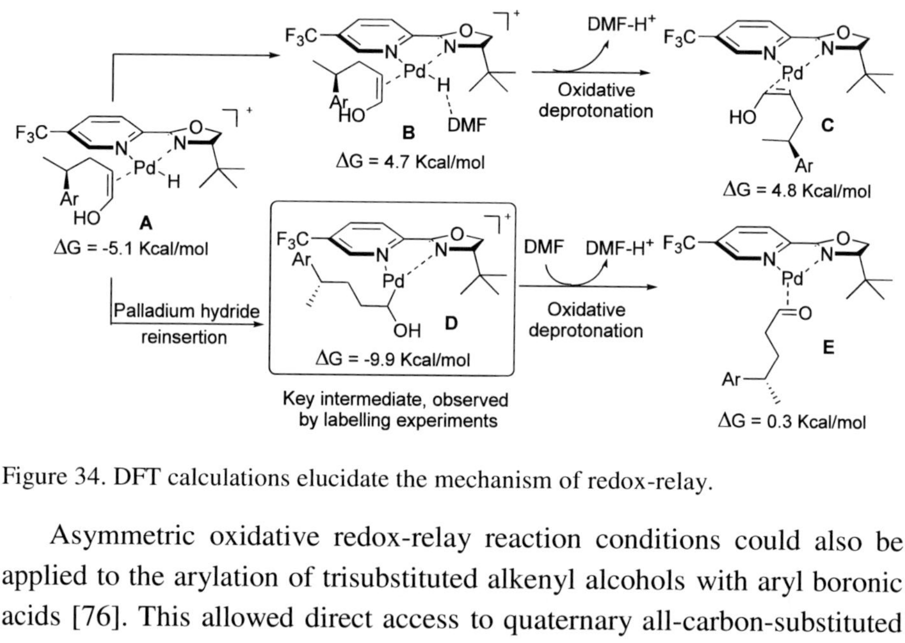

Figure 34. DFT calculations elucidate the mechanism of redox-relay.

Asymmetric oxidative redox-relay reaction conditions could also be applied to the arylation of trisubstituted alkenyl alcohols with aryl boronic acids [76]. This allowed direct access to quaternary all-carbon-substituted β-, γ-, δ-, ε- and ζ-aryl carbonyl compounds, in a highly enantioselective manner. Remote control of enantioselectivity in the formation of all-carbon stereocentres is frequently a synthetic challenge, but of great value, since these structural motifs are frequently found in many secondary metabolites and in some pharmaceutical agents. The conditions required in this case were similar to those previously used by the Sigman group to obtain less substituted substrates (Figure 35). The fact that in cases in which there was a pre-existing stereocentre its stereochemistry was preserved, supports the idea that the catalyst does not become detached from the substrate during migration along the alkyl chain. It is also consistent with the postulation that the catalyst remains ligated to the same face of the alkene during the relay of this unsaturation to the alcohol, which results in the formation of the carbonyl group.

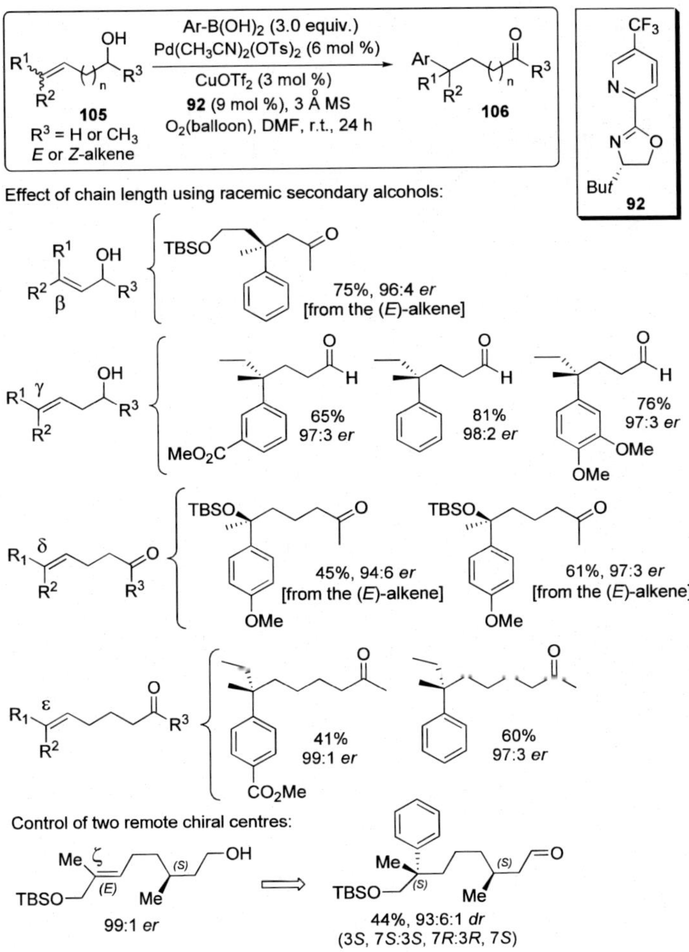

Figure 35. Oxidative redox-relay arylation of trisubstituted alkenyl alcohols.

Competition experiments were performed to clarify further aspects of this reaction. When two alcohols with differing chain lengths were reacted jointly, it was observed that the catalyst prefers to migrate toward the closest alcohol [77]. In addition, an increase in the electron-density in the arene boronic acid was found to promote a faster reaction, which correlates well with Hammett σ_p values.

In 2015 the Sigman group expanded the redox-relay methodologies to the palladium-catalyzed enantioselective Heck alkenylation of alkenols [78]. Until then only aryl boronic acids add been utilized as the alcohol reaction partners. The capability to introduce alkenyl units in an intermolecular fashion is of great interest in synthesis, since it is one of the strategies used to build-up molecular complexity. A method which achieves this in a stereocontrolled fashion away from a carbonyl group, has great potential for applications in the synthesis of natural products and pharmaceutical substances. Despite the simplicity of Heck reactions, up to this stage intermolecular alkenylations were limited to the use of classical cyclic enol ether substrates. However, to install an alkenyl group is different from installing an aryl group: the first reaction goes *via* the formation of a more stable diene intermediate when compared to a styrene intermediate in the second and of a π-allyl intermediate species compared to a π-benzyl intermediate. In addition, Pd(II)-alkenyl intermediates are highly electrophilic. For this procedure to be successful, the hydricity (nucleophilicity) of HA, in intermediate **109** (Figure 36) needs to be reduced, to avoid catalyst arrest arising from the formation of a π-allyl intermediate **111**, *via* β-hydride elimination of HA. The way found to achieve this was to use electron deficient alkenyl triflates derived from β-dicarbonyl compounds. Due to the electron withdrawing nature of the resulting intermediate, the formation of π-allyl intermediates is avoided. Using alkenyl triflates derived from β-dicarbonyl compounds has other attractions, since they are easy to synthesize, for this may be done in a modular fashion [79]. It was found that when these substances were used as substrates, selective β-hydride elimination occurred to give an intermediate of type **112**,

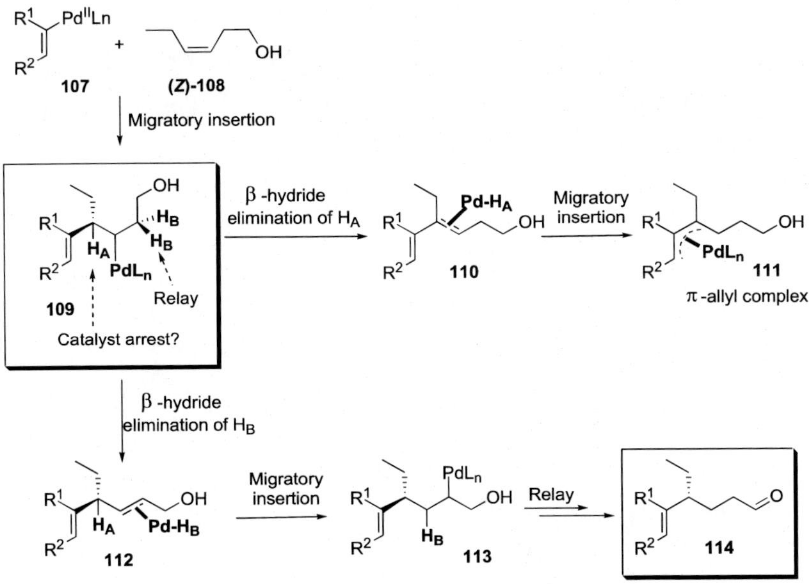

Figure 36. Pd-catalyzed oxidative redox relay Heck alkenylation of alcohols.

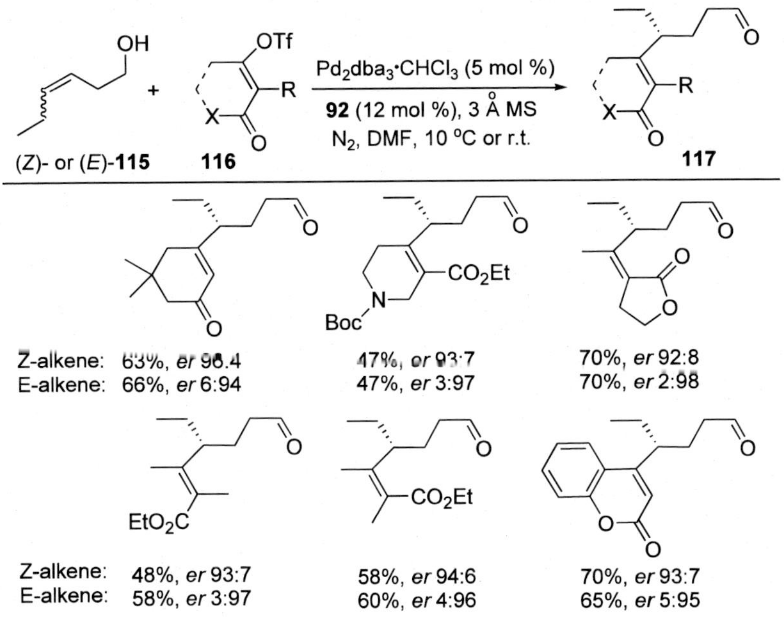

Figure 37. Enantioselective synthesis of β-arylated esters from alcohols.

followed by migration of the catalyst through the alkyl chain to afford the desired alkenylated carbonyl products **114**. Application of this methodology to (*Z*)- or (*E*)-alkenols **115** yielded γ-substituted aldehydes **117** in moderate to good yields and very high *ers* (Figure 37). The method developed was subsequently applied to the synthesis of a tricyclic motif found to be present in various natural products, e.g., elizabethin A or colombiasin A.

Metal-metal catalyst combinations were also employed by Kobayashi and co-workers in TOP processes. A "one-pot" sequential process, consisting of an aerobic oxidation-olefination domino reaction, followed by an asymmetric 1,4-conjugate addition was developed in 2015, which allowed the enantioselective synthesis of enantioenriched β-arylated esters from alcohols, stabilized phosphonates and arylboronic acids [80]. This was the first report on a direct olefination of alcohols through an aerobic oxidation/Horner-Wadsworth-Emmons reaction. Two different types of nanoparticles were used as catalysts. Gold-palladium bimetallic nanoparticles catalysed the domino process and rhodium nanoparticles, i.e., PI/CB-Rh NPs, the third reaction in the sequence (Figure 38).

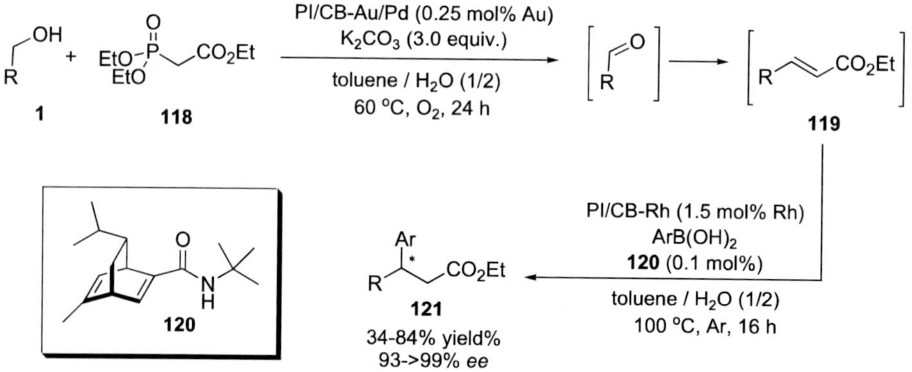

Figure 38. Aerobic oxidation–olefination–asymmetric C–C Bond Formation for the enantioselective synthesis of chiral β-arylated esters.

The Au-Pd NPs consisted of alloys incarcerated in a polystyrene-derived polymer having cross-linking moieties, with carbon black (CB) having a high specific surface area as secondary support (PI/CB–Au/Pd), and were similar to those described previously by the group [40]. The high surface area of the secondary support enhances catalytic activity. An Au/Pd combination provided more robust catalysts having higher oxidation ability than when Au was used alone. Catalyst loading was kept low (0.5 mol % Au) to avoid overoxidation of the newly formed aldehyde to the acid. However, the loading was even lowered further (to 0.25 mol %) when the domino process was combined with the conjugate addition reaction, to avoid an increase in volume of the solid catalysts which reduced stirring efficiency. A lower loading also prevented partial decomposition of the arylboronic acids in the presence of the Au/Pd NPs. The conjugate addition reaction was catalyzed by Rh NPs prepared in a similar fashion which became chirally modified *in situ*, upon co-addition of a chiral diene to the reaction vessel. The reactions were conducted in a toluene/water biphasic system. The desired chiral esters **121**, which are useful as intermediates for the synthesis of various pharmaceutical products and bioactive compounds, could be obtained in high yields and excellent *ees* (93->99%) directly from readily available stable alcohols by this sequential process.

4. Cascade Reactions Promoted by Organocatalysts Only

Despite the fast development of organocatalysis in recent years, the number of enantioselective multicatalysed cascade reactions involving alcohol oxidation is still very limited. In 2013, Khan and Saxena reported a procedure to synthesize complex amines based on the reductive amination of alcohols [81]. A sequential procedure involving alcohol oxidation, followed by imine–iminium formation / reduction domino process was developed. The alcohol oxidation was achieved with catalytic amounts of TEMPO and [bis(acetoxy)iodo]-benzene (BAIB), used as secondary oxidant. This type of oxidation, first developed by Piancatelli and Margarita [82], is particularly useful for "difficult" alcohols. It also has the advantage over some oxidation processes that the reaction by-product is iodobenzene which, being rather inert, does need not be removed from the reaction medium before many subsequent reactions are performed and it is, therefore, rather useful for cascade processes [19]. Another by-product is acetic acid, which may have a beneficial effect on the subsequent reactions. *N,N´*-dimethylpropyleneurea (DMPU) was useful as solvent. Once the oxidation was complete, additional BAIB was added to the reaction vessel together with an amine, Hantszch ester **123** and chiral phosphoric acid **124** as Bronsted acid, upon which an imine–iminium formation/reduction domino reaction took place *in situ*. Either chiral secondary or tertiary amines and octahydropyrazinopyridoindoles could be obtained in this fashion as well as higher ring analogues, starting from simple achiral or racemic alcohols and either primary or secondary amines, respectively, in very high yields and excellent stereoselectivities (Figure 39). Optically active alcohols with a stereocentre at the β-position could also be used as alkylating agents. The configuration of the stereocentre was preserved, contrary to what has been observed with ruthenium catalyzed methods, in which epimerization of such stereocentres occurs. The alkylation of optically active amines bearing a stereocentre at the α-position could also be achieved in a highly stereoselective fashion. Again, the existing stereocentre was not affected by the reaction conditions. The procedure was also applied to intramolecular reactions and amino alcohols could be cyclized into 2° or 3° amines.

This sequential cascade allowed the synthesis in high *ees* of important tetracyclic scaffolds, i.e., octahydropyrazinopyridoindole derivatives, by coupling chiral non-racemic tetrahydropyridoindoles with *N*-protected amino alcohols possessing various carbon chain lengths.

Schreiner and co-workers expanded the concept of bifunctional catalysis, illustrated in Figure 2, to the development of multi-functional, peptide-based catalysts. In contrast with typical bifunctional catalysts like thioureas or phosphoric acids, which can activate one or more substrates simultaneously, in different ways, *via* different functional groups present in the molecule, the new catalysts incorporate in their structure components which can catalyse different reactions in a domino fashion. For example, catalyst **126**, incorporating a TEMPO oxidizer and a π-Me-histidine residue, kept apart in the molecule by a spacer, i.e., compartmentalized, so that their functions do not interfere with each other, was used to catalyse the one-pot desymmetrization/oxidation of meso-1,2-alkane diols **127** (Figure 40) [83]. In these reactions, which were performed in a sequential fashion, *m*-chloroperbenzoic acid (MCPBA) was employed as the terminal oxidant. MCPBA oxidizes the nitroxide i.e., TEMPO, into an oxoammonium salt, which is capable of delivering a carbonyl compound upon reaction with the alcohol [19].

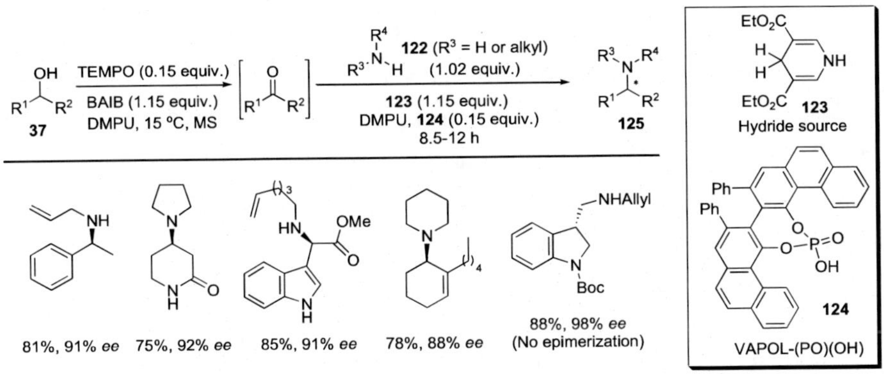

Figure 39. "One-pot" oxidation/imine–iminium formation/ reduction cascade for the synthesis of chiral amines, *via* TEMPO–BAIB–HEH–Brønsted acid catalysis.

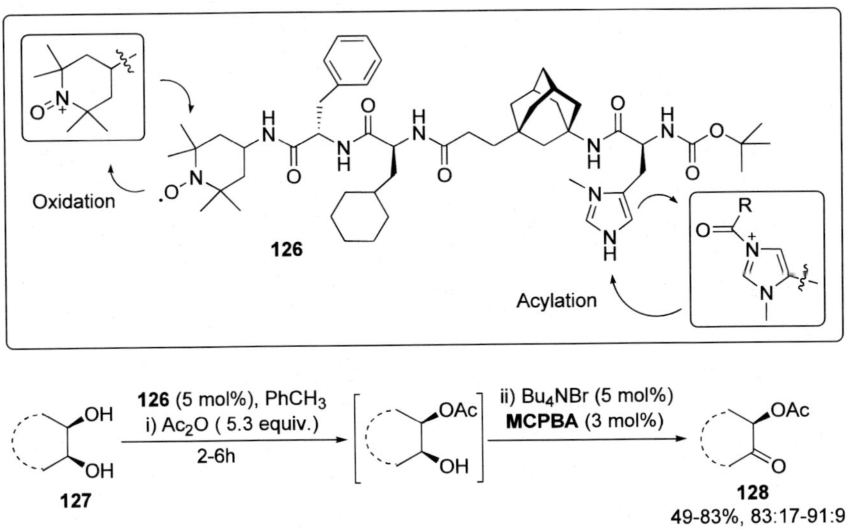

Figure 40. Kinetic resolution of diols *via* oxidative esterification using multifunctional catalyst **126**.

In 2015, the group reported that the same catalyst could be used for alcohol cross-coupling for the kinetic resolution of diols *via* oxidative esterification [84]. This process is discussed further in the chapter on kinetic resolution.

5. Bioorgano Catalysis

The ability to imitate Nature and use its catalysts, i.e., enzymes or whole microorganisms, which are usually non-toxic, biodegradable, biocompatible and operate under physiological conditions, is particularly appealing, if one desires to practice sustainable chemistry. It is useful when considering safety and economic benefits too. However, although multistep enzymatic reactions in a wholecell have been performed since the late 1980s, e.g., to produce amino acids [85], and several enzyme classes have also been combined successfully in one-pot processes to afford complex cascade reactions [86], the combination of enzymes with either organocatalysts or metal catalysts, besides the odd example, is a far more recent development and a field still very much in its infancy [87-89]. The first two types of catalysis will not be covered in this chapter, since a large body of information is available which would warrant a chapter on its own. The interested reader is referred to the recent reviews on these topics for further information [30-32, 85, 86].

A few methods of synthesis have been developed using multicatalytic chemoenzymatic approaches involving alcohol oxidation. One such hybrid system was developed by Galli and co-workers in 2001. It combines a multi-copper oxidase, a laccase, with the radical organocatalyst TEMPO and it has proved to be useful for the oxidation of a variety of alcohols [90, 91]. It is thought that TEMPO oxidizes the alcohol and transfers the electrons to the laccase, which in turn reduces oxygen. This system is known to oxidize primary and secondary alcohols, diols, isosorbide and isomannide [91]. The laccase-TEMPO system has been utilized in a few cascade reactions, including enantioselective processes. In particular, useful cascades have been developed to produce enantio pure amines and alcohols, through combinations of non-enantioselective oxidation of racemic secondary alcohols and highly enantioselective ketone to amine or alcohol conversion. These substances are often of interest as starting materials for the pharmaceutical industry. In one such example, described by Gotor-Fernández, Lavandera and co-workers, is a sequential procedure in which the laccase/TEMPO hybrid system oxidizes alcohols **129** into ketones which are subsequently transformed, in "one-pot," into optically enriched amines **130** (in 90-99% *ee*) through the action of transaminases [92]. The transaminases used were pyridoxal 5'-phosphate (PLP)-dependent enzymes. Either *S*- or *R*-configured amines could be obtained, depending on the selectivity of the transaminase employed. Conversions were in the range 67-99%, with the remaining reaction products obtained being the starting alcohols (<1%) and intermediate ketones. For the sequential procedure to work well the pH add to be adjusted between reactions, due to differences in the optimum working pH of the enzymes employed. The presence of isopropyl amine was essential for success. It was used-up in the reaction as a sacrificial amine donor (Figure 41).

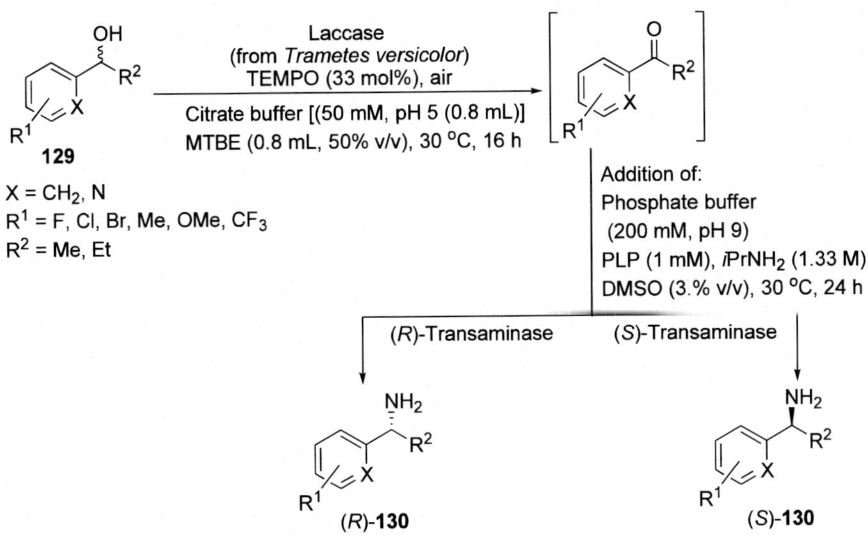

Figure 41. Chemoenzymatic oxidative amination of racemic *sec*-alcohols.

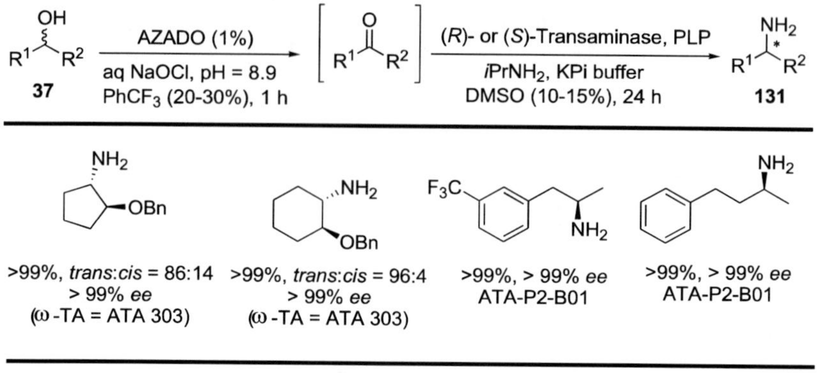

Figure 42. Hybrid bioorgano catalysis for the asymmetric amination of alcohols.

Also in 2017 Rebolledo, González-Sabín and co-workers showed that there are other possibilities for the conversion of racemic secondary alcohols into enantioenriched amines. In this case the organocatalyst was 2-azaadamantane N-oxyl (AZADO) combined with NaOCl as oxidant and the enzyme ω-transaminase, for a one-pot oxidation–transamination sequential process in aqueous medium [93]. The choice of organocatalyst was motivated by previous reports which revealed that AZADO displayed a catalytic proficiency superior to that of TEMPO, which made possible the oxidation of structurally hindered secondary alcohols under a variety of conditions [94]. In fact it was found that 1% of AZADO was enough for the alcohol to ketone conversion to be achieved when NaOCl was used as terminal oxidant. The procedure developed could even be applied to sterically hindered β-substituted cycloalkanols, where a highly stereoselective dynamic asymmetric amination yielded products **131** with two contiguous stereocenters with very high enantio- and diastereomeric ratio (>90% yield, >99% *ee*, and up to 49:1 *dr*) (Figure 42).

6. Biometal Catalysis

Enzyme-metal combinations are often limited by solubility issues, since the majority of metal catalysts cannot be utilized in aqueous systems, the ideal media for enzymes. The requirement of many metals for the exclusion of oxygen and operation under inert atmospheres imposes further restrictions. Notable exceptions are dynamic kinetic resolutions, which are particularly successful for secondary alcohols, diols, amines and esters [95, 96]. Special conditions for compatibility have, nevertheless, to be found, since the metal may even react with the product of the enzymatic resolution or interfere with it, and the enzyme, in turn, together with its accompanying additives (e.g., surfactants and stabilizers), may inhibit the reaction of the metal [96, 97]. DKRs were first shown to be possible for allylic acetates [17] and secondary alcohols [98] using metal-enzyme combinations, by Williams and co-workers in 1996. Soon after a few groups, independently, illustrated how metal-enzyme combinations could be used to effect tandem, *in situ*, racemization and resolution, Reetz [99], Stürmer [100] and Bäckvall [101], which extended the scope of DKRs. Chemoenzymatic DKRs typically involve two basic processes, one in which the enzyme acts as an enantioselective resolving catalyst and another in which the metal works as the racemising catalyst. Ruthenium is the most used metal, but others have also been shown to be useful, e.g., palladium, iridium, nickel, zinc, depending on the substrates and on the reactions involved. In the William's DKR of secondary alcohols, the alcohol is temporarily oxidized and subsequently reduced, with one enantiomer being preferentially acylated with a lipase from *P. fluorescens* (Figure 43). A rhodium catalyst performed the best in this process, in which addition of extra acetophenone facilitates hydrogen transfer [98].

The racemization mechanism of DKRs depends on the substrate and on the metal catalyst employed. Typical mechanisms are reversible hydrogen transfer processes, outer-sphere redox mechanisms and dehydrative mechanisms. In the case of ruthenium-catalyzed DKR of allylic alcohols, isomerization is involved. Due to the large volume of information, DKRs involving hydrogen transfer processes (temporary oxidation, followed by reduction), would warrant a separate chapter and will not be discussed further here. The reader is referred to the recent reviews already cited for further information. Very few examples involve simple alcohol oxidation. One such example was reported in 2010. Kroutil and co-workers were able to achieve, in one-pot, a domino iridium catalysed oxidation/enzymatic reduction employing orthogonal reagents, showing for the first time that iridium and alcohol dehydrogenase-catalyzed redox reactions are compatible [103]. The procedure allowed the conversion of racemic chlorohydrins (**134**) into enantioenriched ones in >99.9% yield and good *ees* of up to 40% (Figure 44).

The success of this methodology was also dependent on the use of an additional ketone as hydrogen acceptor, for which cyclohexanone **136** proved to be the best, as well as of a third catalyst, formate dehydrogenase (FDH). The role of the third catalyst was to be an orthogonal reducing agent for the biocatalytic reduction. This was necessary, since 2-propanol acted as a hydrogen donor for both the enzyme and the metal catalyst. The combination of formate and FDH worked well, because when tested in an additional experiment, it was observed that the metal catalyst did not accept formate. Hence it could be used for nicotinamide adenine dinucleotide (NADH) recycling.

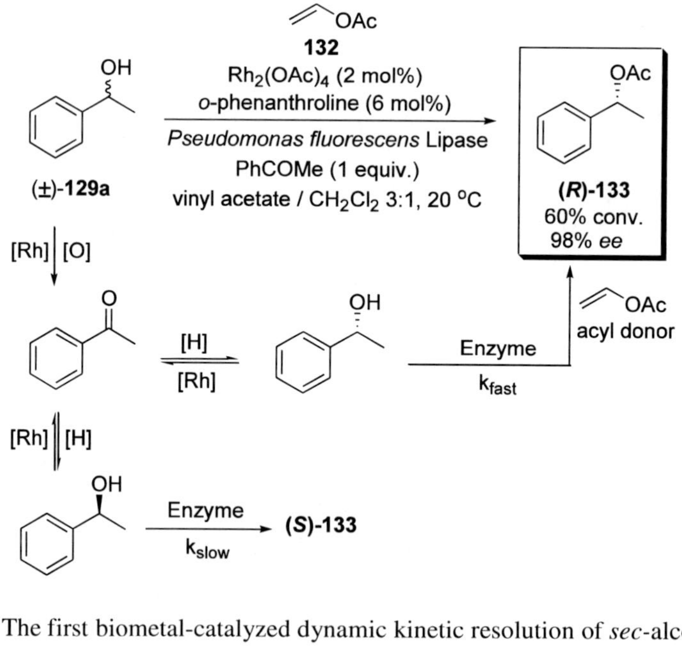

Figure 43. The first biometal-catalyzed dynamic kinetic resolution of *sec*-alcohols.

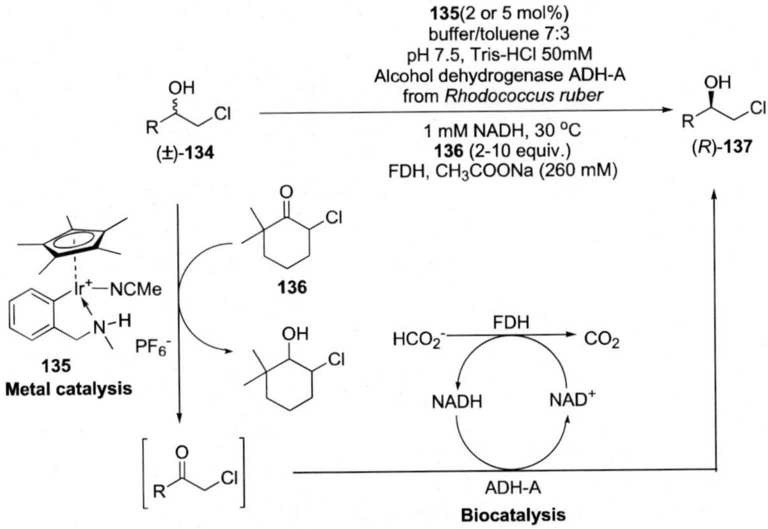

Figure 44. Iridium catalysed oxidation/enzymatic reduction for the DKR of *sec*-alcohols - **134**.

CONCLUSION

Although the application of enantioselective multicatalytic approaches in synthesis is still a young science, the examples available clearly indicate the enormous potential for future development. Not only do these systems allow transformations not accessible with one catalyst only but, being "one-pot approaches," they are more economical, less time consuming, lower waste generating, i.e., they are more efficient and more sustainable. Some challenges arise, besides those of compatibility issues between all the reagents mixed together: the control of minute impurities which can be carried over if the processes are to be used for the manufacture of pharmaceuticals, unfavourable equilibria or unstable (highly reactive) products and intermediates, the need to overcome the high temperatures which are generally required with some of the methodologies, e.g., in hydrogen borrowing, which may be detrimental to enantioselectivity. The development of efficient hybrid catalysts, even heterogeneous, which may be easily applied to industrial processes or others for continuous flow synthesis appear favourable, in face of the developments seen so far. Enzyme engineering may bring additional benefits as well as mechanistic/theoretical studies to help to understand the processes involved. Seeing the fast pace at which this field is developing, it is expectable to see many interesting new developements in the near future.

ACKNOWLEDGMENTS

Financial support from Fundação para a Ciência e a Tecnologia (FCT), Portugal, in the form of projects PTDC/QEQ-QIN/3967/2014, UID/QUI/00100/2013, IF/01270/2013/CP1163/CT0007, and "Investigador 2013" Program (to M. N. Kopylovich), is gratefully acknowledged.

REFERENCES

[1] Li, C. J. and Trost, B. M. 2008. Green chemistry for chemical synthesis. *Proc. Nat. Acad. Sci. US*, 105 (36): 13197-13202.

[2] Blum, C., Bunke, D., Hungsberg, M., Roelofs, E., Joas, A., Joas, R., Blepp, M., Stolzenberg, H. C. 2017. The concept of sustainable chemistry: Key drivers for the transition towards sustainable development. *Sustainable Chem. Pharm.*, 5: 94-104.

[3] Lanzafame, P., Perathoner, S., Centi, G., Gross, S. and Hensen, E. J. M. 2017. Grand challenges for catalysis in the Science and Technology Roadmap on Catalysis for Europe: moving ahead for a sustainable future. *Catal. Sci. Technol.*, 7 (22): 5182-5194.

[4] Xu, P. F. and W. Wang, eds. 2014. *Catalytic Cascade Reactions*. New Jersey: John Wiley & Sons.

[5] Inamdar, S. M., Shinde, V. S. and Patil, N. T. 2015. Enantioselective cooperative catalysis, *Org. Biomol. Chem.*, 13 (30): 8116-8162.

[6] Chauhan, P., Mahajan, S. and Enders, D. 2017. Achieving molecular complexity via stereoselective multiple domino reactions promoted by

a secondary amine organocatalyst. *Acc. Chem. Res.*, 50 (11), 2809-2821.

[7] Schmidt, S., Castiglione, K. and Kourist, R. 2018. Overcoming the incompatibility challenge in chemoenzymatic and multi-catalytic cascade reactions. *Chem. Eur. J.*, 24 (8), 1755-1768.

[8] Lu, J., Dimroth, J. and Weck, M. 2015. Compartmentalization of incompatible catalytic transformations for tandem catalysis. *J. Am. Chem. Soc.*, 137 (40), 12984-12989.

[9] Weissermel, K, and Arpe, H. J. 2003. *Industrial organic chemistry book*, 4th edn., Weinheim: Wiley-VCH.

[10] Wu, L., Moteki, T., Gokhale, A. A., Flaherty, D. W. and Toste, F. D. 2016. Production of fuels and chemicals from biomass: condensation reactions and beyond. *Chem.* 1: 32-58.

[11] Grand View Research. 2017. M*arket Research Report ID: GVR-2-68038-236-5.* Accessed July 2018. https://www.grandviewresearch.com/industry-analysis/industrial-alcohol-market.

[12] Faisca Phillips, A. M., Pombeiro, A. J. L., Kopylovich, M. N. 2017. Recent advances in cascade reactions inated by alcohol oxidation, *ChemCatChem,* 9 (2): 217-246.

[13] Laphookhieo, S., Jones, S., Raw, S. A., Sainz, Y. F. and Taylor, R. J. K. 2006. Tandem oxidation processes for the regioselective preparation of 5-substituted and 6-substituted 1,2,4-triazines, *Tetrahedron Lett.* 47 (23): 3865-3870.

[14] Huang, C. C. 1987. Synthesis of carbon-14 labeled 1-(4-methoxybenzoyl)-5-oxo-2-pyrrolidinepropanoic acid (CI-933). *J. Labelled Compd. Radiopharm.*, 24 (6): 675-681.

[15] Taylor, R. J. K., Alcaraz, L., Kapfer-Eyer, I., Macdonald, G., Wei, X. and Lewis, N. J. (1998). The Synthesis of Alisamycin, Nisamycin, LL-C10037*α* and novel epoxyquinol and epoxyquinone analogues of Manumycin A, *Synthesis*, 5: 775-790.

[16] Taylor, R. J. K., Reid, M., Foot, J. and Raw, S. A. (2005). Tandem Oxidation Processes Using Manganese Dioxide: Discovery, Applications, and Current Studies. *Acc. Chem. Res.*, 38: 851-869.

[17] Allen, J. V. and Williams, J. M. J. 1996. Dynamic kinetic resolution with enzyme and palladium combinations. *Tetrahedron Lett.* 37 (11): 1859-1862.

[18] Rueping, M., Sundén, H. and Sugiono, E. (2012). Unifying metal- and organocatalysis for asymmetric oxidative iminium activation: a relay catalytic system enabling the combined allylic oxidation of alcohols and prolinol ether catalyzed iminium reactions. *Chem. Eur. J.*, 18 (12): 3649-3653.

[19] Tojo, G. and Fernández, M. (2006). *Oxidation of Alcohols to Aldehydes and Ketones: A Guide to Current Common Practice*. Edited by G. Tojo. New York: Springer.

[20] Kopylovich, M. N., Ribeiro, A. P. C., Alegria, E. C. B. A., Martins, N. M. R., Martins, L. M. D. R. S. and Pombeiro, A. J. L. 2015. "Catalytic Oxidation of Alcohols: Recent Advances." In *Advances in Organometallic Chemistry,* vol. 63, edited by P. J. Pérez, 91-174. Massachusetts: Elsevier.

[21] Jeena, V. and Robinson, R. S. 2014. Recent developments in one-pot tandem oxidation process coupling reactions. *RSC Adv.*, 4 (77): 40720-40739.

[22] Xu, C., Zhang, C., Li, H., Zhao, X., Song, L. and Li, X. 2016. An overview of selective oxidation of alcohols: catalysts, oxidants and reaction mechanisms. *Catal. Surv. Asia*, 20 (1):13-22.

[23] Gunanathan, C. and Milstein, D. 2013. Applications of acceptorless dehydrogenation and related transformations in chemical synthesis. *Science*, 341 (6143): 249-260.

[24] Siddiki, S. M. A. H., Toyao, T. and Shimizu, K. 2018. Acceptorless dehydrogenative coupling reactions with alcohols over heterogeneous catalysts. *Green Chem., 20 (13):* 2933-2952.

[25] Haniti, M., Hamid, S. A., Slatford, P. A. and Williams, J. M. J. 2007. Borrowing hydrogen in the activation of alcohols. *Adv. Synth. Catal.*, 349 (10): 1555-1575.

[26] Obora, Y. 2014. Recent advances in α-alkylation reactions using alcohols with hydrogen borrowing methodologies. *ACS Catal.*, 4 (11): 3972-3981.

[27] Corma, A., Navas, J. and Sabater, M. J. 2018. Advances in one-pot synthesis through borrowing hydrogen catalysis. *Chem. Rev.*, 118 (4): 1410-1459.

[28] Ketcham, J. M., Shin, I., Montgomery, T. P. and Krische, M. J. 2014. Catalytic enantioselective C–H functionalization of alcohols by redox-triggered carbonyl addition: borrowing hydrogen, returning carbon. *Angew. Chem. Int. Ed.*, 53 (35): 9142-9150.

[29] Kim, S. W., Zhang, W. and Krische, M. J. 2017. Catalytic enantioselective carbonyl allylation and propargylation via alcohol-mediated hydrogen transfer: merging the chemistry of Grignard and Sabatier. *Acc. Chem. Res.*, 50 (9): 2371-2380.

[30] Ricca, E., Brucher, B. and Schrittwieser, J. H. 2011. Multi-enzymatic cascade reactions: overview and perspectives. *Adv. Synth. Catal.*, 353 (13): 2239-2262.

[31] Köhler, V. and Turner, N. J. 2015. Artificial concurrent catalytic processes involving enzymes. *Chem. Commun.*, 51 (3): 450-464.

[32] Sperl, J. M. and Sieber V. 2018. Multienzyme cascade reactions - status and recent advances. *ACS Catal.*, 8 (3): 2385-2396.

[33] Reetz, M. T. 2013. Biocatalysis in organic chemistry and biotechnology: past, present, and future. *J. Am. Chem. Soc.*, 135 (34): 12480-12496.

[34] Akagawa, K., Fujiwara, T., Sakamoto, S. and Kudo, K. 2010. One-pot sequential alcohol oxidation and asymmetric a-oxyamination in aqueous media using recyclable resin-supported peptide catalyst. *Chem. Commun.,* 46 (42): 8040-8042.

[35] Sheldon, R. A., Arends, I. W. C. E., Brink, G. J. T. and Dijksman, A 2002. Green, catalytic oxidations of alcohols. *Acc. Chem. Res.*, 35 (9): 774-781.

[36] Griyth, W. P., Ley, S. V., Whitcombe, G. P. and White, A. D. 1987. Preparation and use of tetra-n-butylammonium per-ruthenate (TBAP reagent) and tetra-n-propylammonium per-ruthenate (TPAP reagent) as new catalytic oxidants for alcohols. *J. Chem. Soc., Chem. Commun.*, 21: 1625-1627.

[37] Rueping, M., Sundén, H., Hubener, L. and Sugiono, E. 2012. Asymmetric oxidative Lewis base catalysis - unifying iminium and enamine organocatalysis with oxidations. *Chem. Commun.*, 48 (16): 2201-2203.

[38] Rueping, M., Dufour, J. and Maji, M. S. 2012. Relay catalysis: combined metal catalyzed oxidation and asymmetric iminium catalysis for the synthesis of bi- and tricyclic chromenes. *Chem. Commun.*, 48 (28): 3406-3408.

[39] Rueping, M., Dufour, J. and Bui, L. 2014. Convergent catalysis: asymmetric synthesis of dihydroquinolines using a combined metal catalysis and organocatalysis approach. *ACS Catal.*, 4 (3): 1021-1025.

[40] Miyamura, H., Choo, G. C. Y., Yasukawa, T., Yoo, W. J. and Kobayashi, S. 2013. A heterogeneous layered bifunctional catalyst for the integration of aerobic oxidation and asymmetric C–C bond formation. *Chem. Commun.*, 49 (85): 9917-9919.

[41] Yoo, W. J., Miyamura, H. and Kobayashi, S. 2011. Polymer-incarcerated gold–palladium nanoclusters with boron on carbon: a mild and efficient catalyst for the sequential aerobic oxidation–Michael addition of 1,3-dicarbonyl compounds to allylic alcohols. *J. Am. Chem. Soc.*, 133 (9): 3095-3103.

[42] Rana, N. K., Joshi, H., Jha, R. K. and Singh, V. K. 2017. Enantioselective tandem oxidation/Michael–aldol approaches to tetrasubstituted cyclohexanes. *Chem. Eur. J.*, 23 (9): 2040-2043.

[43] Hamid, M. H. S. A., Slatford, P. A. and Williams, J. M. J. 2007. Borrowing hydrogen in the activation of alcohols. *Adv. Synth. Catal.*, 349 (10): 1555-1575.

[44] Hollmann, D. 2014. Advances in asymmetric borrowing hydrogen catalysis. *ChemSusChem*, 7 (9): 2411-2413.

[45] Shimizu, K. I. 2015. Heterogeneous catalysis for the direct synthesis of chemicals by borrowing hydrogen methodology. *Catal. Sci. Technol.*, 5 (3): 1412-1427.

[46] Shermer, D. J., Slatford, P. A., Edney, D. D. and Williams, J. M. J. 2007. Borrowing hydrogen in an indirect asymmetric Wittig reaction. *Tetrahedron: Asymmetry*, 18 (24): 2845-2848.

[47] Quintard, A., Constantieux, T. and Rodriguez, J. 2013. An iron/amine-catalyzed cascade process for the enantioselective functionalization of allylic alcohols. *Angew. Chem. Int. Ed.*, 52 (49): 12883-12887.

[48] Roudier, M., Constantieux, T., Rodriguez, J. and Quintard, A. 2016. Recent achievements in enantioselective borrowing hydrogen by the combination of iron- and organocatalysis. *Chimia*, 70 (1/2): 97-101.

[49] Knölker, H. J., Heber, J. and Mahler, C. H. 1992. *Synlett*, 12: 1002-1004.

[50] Roudier, M., Constantieux, T., Quintard, A. and Rodriguez, J. 2014. Enantioselective cascade formal reductive insertion of allylic alcohols into the C(O)–C bond of 1,3-diketones: ready access to synthetically valuable 3-alkylpentanol units. *Org. Lett.*, 16 (11): 2802-2805.

[51] Brenna, E., Fuganti, C., Ronzani, S. and Serra, S. 2002. Biocatalyzed preparation of the optically enriched stereoisomers of 4-methyl-2-phenyl-tetrahydro-2*H*-pyran (Doremox®) *Can. J. Chem.*, 80 (6): 714-723.

[52] Zhang, Y., Lim, C. S., Sim, D. S. B., Pan, H. J. and Zhao, Y. 2014. Catalytic enantioselective amination of alcohols by the use of borrowing hydrogen methodology: cooperative catalysis by iridium and a chiral phosphoric acid. *Angew. Chem. Int. Ed.*, 53 (5): 1399-1403.

[53] Rong, Z. Q., Zhang, Y., Chua, R. H. B., Pan, H. J. and Zhao, Y. 2015. Dynamic kinetic asymmetric amination of alcohols: from a mixture of four isomers to diastereo- and enantiopure α-branched amines. *J. Am. Chem. Soc.*, 137 (15): 4944-4947.

[54] Lim, C. S., Quach, T. T. and Zhao, Y. 2017. Enantioselective synthesis of tetrahydroquinolines by borrowing hydrogen methodology: cooperative catalysis by an achiral iridacycle and a chiral phosphoric acid. *Angew. Chem. Int. Ed.*, 56 (25), 7176-7180.

[55] Akagawa, K., Takigawa, S., Mano, E. and Kudo, K. 2011. Sequential oxidation/asymmetric aldol reaction of primary alcohols by resin-supported catalysts. *Tetrahedron Lett.*, 52 (7): 770-773.

[56] Akagawa, K., Sakamoto, S. and Kudo, K. 2007. Resin-supported acid- and base-catalyzed one-pot sequential reaction including an enantioselective step. *Tetrahedron Lett.*, 48 (6): 985-987.

[57] Ho, X. H., Oh, H. J. and Jang, H. Y. 2012. Multistep organocatalysis for the asymmetric synthesis of multisubstituted aldehydes from allylic alcohols. *Eur. J. Org. Chem.*, 29: 5655-5659.

[58] Deiana, L., Jiang, Y., Palo-Nieto, C., Afewerki, S., Incerti-Pradillos, C. A., Verho, O., Tai, C. W., Johnston, E. V. and Córdova, A. 2014. Combined heterogeneous metal/chiral amine: multiple relay catalysis for versatile eco-friendly synthesis. *Angew. Chem. Int. Ed.*, 53 (13): 3447-3451.

[59] Luca Deiana, Lorenza Ghisu, Oscar Córdova, Samson Afewerki, Renyun Zhang, Armando Córdova. 2014. Efficient and highly enantioselective aerobic oxidation–Michael–carbocyclization cascade transformations by integrated Pd(0)-CPG nanoparticle/ chiral amine relay catalysis integrated heterogeneous Pd(0)-CPG nanoparticle catalysts. *Synthesis*, 46 (10): 1303-1310.

[60] Córdova, A. 2015. *Pure Appl. Chem.*, 87 (8): 1011-1019.

[61] Shakeri, M., Tai, C. W., Göthelid, E., Oscarsson, S. and Bäckvall, J. E. 2011. Small Pd nanoparticles supported in large pores of mesocellular foam: an excellent catalyst for racemization of amines. *Chem. Eur. J.*, 17 (47): 13269-13272.

[62] Johnston, E. V., Verho, O., Kärkäs, M. D., Shakeri, M., Tai, C. W., Palmgren, P., Eriksson, K., Oscarsson, S. and Bäckvall, J. E. 2012. Highly dispersed palladium nanoparticles on mesocellular foam: an efficient and recyclable heterogeneous catalyst for alcohol oxidation. *Chem. Eur. J.*, 18 (39): 12202-12206.

[63] Santoro, S., Deiana, L., Zhao, G. L., Lin, S., Himo, F. and Córdova, A. 2014. Mechanism of palladium/amine cocatalyzed carbocyclization of aldehydes with alkynes and its merging with "Pd oxidase catalysis." *ACS Catal.*, 4 (12): 4474-4484.

[64] Stahl, S. S. 2004. Palladium oxidase catalysis: selective oxidation of organic chemicals by direct dioxygen-coupled turnover. *Angew. Chem. Int. Ed. Engl.*, 43 (26): 3400-3420.

[65] Parmar, D., Sugiono, E., Raja, S. and Rueping, M. 2014. Complete field guide to asymmetric BINOL-phosphate derived Brønsted acid and metal catalysis: history and classification by mode of activation; Brønsted acidity, hydrogen bonding, ion pairing, and metal phosphates. *Chem. Rev.*, 114 (18): 9047-9153.

[66] Faisca Phillips, A. M. and Pombeiro, A. J. L. 2017. Recent advances in organocatalytic enantioselective transfer hydrogenation. *Org. Biomol. Chem.*, 15 (11): 2307-2340.

[67] Cheng, H. G., Miguélez, J., Miyamura, H., Yoo, W. J. and Kobayashi, S. 2017. Integration of aerobic oxidation and intramolecular asymmetric aza-Friedel–Crafts reactions with a chiral bifunctional heterogeneous catalyst. *Chem. Sci.*, 8 (2): 1356-1359.

[68] Quintard, A., Roudier, M. and Rodriguez, J. 2018. Multicatalytic enantioselective borrowing hydrogen δ-lactonization strategy from β-keto esters and allylic alcohols. *Synthesis*, 50 (4): 785-792.

[69] See, for example, Fan, Y. Y., Zhang, H., Zhou, Y., Liu, H. B., Tang, W., Zhou, B., Zuo, J. P. and Yue, J. M. 2015. Phainanoids A–F, a new class of potent immunosuppressive triterpenoids with an unprecedented carbon skeleton from *Phyllanthus hainanensis*. *J. Am. Chem. Soc.*, 137 (1): 138-141.

[70] Werner, E. W., Mei, T. S., Burckle, A. J. and Sigman, M. S. 2012. Enantioselective Heck arylations of acyclic alkenyl alcohols using a redox-relay strategy. *Science*, 338 (6113): 1455-1458.

[71] Lee, A. L. 2016. Enantioselective oxidative boron Heck reactions. *Org. Biomol. Chem.*, 14 (24): 5357-5366.

[72] Xu, L., Hilton, M. J., Zhang, X., Norrby, P., Wu, Y. D., Sigman, M S. and Wiest, O. 2014. Mechanism, reactivity and selectivity in palladium-catalyzed redox-relay Heck arylations of alkenyl alcohols. *J. Am. Chem. Soc.*, 136 (5): 1960-1967.

[73] Hilton, M. J., Xu, L. P., Norrby, P., Wu, Y. D., Wiest, O. and Sigman, M. S. 2014. Investigating the nature of palladium chain-walking in the enantioselective redox-relay Heck reaction of alkenyl alcohols. *J. Org. Chem.*, 79 (24): 11841-11850.

[74] Mei, T. S., Werner, E. W., Burckle, A. J. and Sigman, M. S. 2013. Enantioselective redox-relay oxidative Heck arylations of acyclic alkenyl alcohols using boronic acids. *J. Am. Chem. Soc.*, 135 (18): 6830-6833.

[75] Dang, Y., Qu, S., Wang, Z. X. and Wang, X. 2014. A computational mechanistic study of an unprecedented Heck-type relay reaction: insight into the origins of regio- and enantioselectivities. *J. Am. Chem. Soc.*, 136 (3), 986-998.

[76] Mei, T. S., Patel, H. H. and Sigman, M. S. 2014. Enantioselective construction of remote quaternary centres. *Nature*, 508 (7496): 340-344.

[77] Hilton, M. J., Cheng, B., Buckley, B. R., Xu, L., Wiest, O. and Sigman, M. S. 2015. Relative reactivity of alkenyl alcohols in the palladium-catalyzed redox-relay Heck reaction. *Tetrahedron*, 71 (37): 6513-6518.

[78] Patel, H. H. and Sigman, M. S. 2015. Palladium-catalyzed enantioselective Heck alkenylation of acyclic alkenols using a redox-relay strategy. *J. Am. Chem. Soc.* 137 (10): 3462-3465.

[79] Chassaing, S., Specklin, S., Weibel, J. M. and Pale, P. 2012. Vinyl triflates derived from 1,3-dicarbonyl compounds and analogs: access and applications to organic synthesis. *Tetrahedron*, 68 (36):7245-7273.

[80] Miyamura, H., Suzuki, A., Yasukawa, T. and Kobayashi, S. 2015. Integrated process of aerobic oxidation–olefination–asymmetric C–C bond formation catalyzed by robust heterogeneous gold/palladium and chirally modified rhodium nanoparticles. *Adv. Synth. Catal.*, 357 (18): 3815-3819.

[81] Khan, I. A. and Saxena, A. K. 2013. Metal-free, mild, nonepimerizing, chemo- and enantio- or diastereoselective *N*-alkylation of amines by alcohols via oxidation/imine–iminium formation/reductive amination: a pragmatic synthesis of octahydropyrazinopyridoindoles and higher ring analogues, *J. Am. Chem. Soc.*, 78 (23): 11656-11669.

[82] De Mico, A., Margarita, R., Parlanti, L., Vescovi, A. and Piancatelli, G. 1997. A versatile and highly selective hypervalent iodine

(III)/2,2,6,6-tetramethyl-1-piperidinyloxyl-mediated oxidation of alcohols to carbonyl compounds, *J. Org. Chem.*, 62 (20): 6974-6977.

[83] Müller, C. E., Hrdina, R., Wende, R. C. and Schreiner, P. R. 2011. A multicatalyst system for the one-pot desymmetrization/oxidation of meso-1,2-alkane diols. *Chem. Eur. J.*, 17 (23): 6309-6314.

[84] Hofmann, C., Schümann, J. M. and Schreiner, P. R. 2015. Alcohol cross-coupling for the kinetic resolution of diols via oxidative esterification. *J. Org. Chem.*, 80 (3): 1972-1978.

[85] Wu, S. and Li, Z. 2017. Whole-cell cascade biotransformations for one-pot multistep organic synthesis. *ChemCatChem*, 10 (10), 2164-2178.

[86] Schrittwieser, J. H., Velikogne, S., Hall, M. and Kroutil, W. 2018. Artificial biocatalytic linear cascades for preparation of organic molecules, *Chem. Rev.*, *118* (1): 270-348.

[87] Wang, Y. and Zhao, H. 2016. Tandem reactions combining biocatalysts and chemical catalysts for asymmetric synthesis. *Catalysts*, 6 (12): 194.

[88] Schmidt, S., Castiglione, K. and Kourist, R. 2018. Overcoming the incompatibility challenge in chemoenzymatic and multi-catalytic cascade reactions. *Chem. Eur. J.*, 24 (8): 1755-1768.

[89] Köhler, V. and Turner, N. J. 2015. Artificial concurrent catalytic processes involving enzymes. *Chem. Commun.*, 51 (3), 450-464.

[90] Fabbrini, M., Galli, C., Gentili, P. and Macchitella, D. 2001. An oxidation of alcohols by oxygen with the enzyme laccase and mediation by TEMPO. *Tetrahedron Lett.*, 42 (43): 7551-7553.

[91] Liu, J., Wu, S. and Li, Z. 2018. Recent advances in enzymatic oxidation of alcohols, *Curr. Opin. Chem. Biol.*, 43: 77-86.

[92] Martínez-Montero, L., Gotor-Fernández, V. and Lavandera, I. 2017. Stereoselective amination of racemic sec-alcohols through sequential application of laccases and transaminases. *Green Chem.*, 19 (2): 474-480.

[93] Liardo, E., Ríos-Lombardía, N., Morís, F., Rebolledo, F. and González-Sabín, J. 2017. Hybrid organo- and biocatalytic process for

the asymmetric transformation of alcohols into amines in aqueous medium. *ACS Catal.*, 7 (7): 4768-4774.

[94] Shibuya, M., Tomizawa, M. T., Suzuki, I. and Iwabuchi, Y. 2006. 2-Azaadamantane-*N*-Oxyl (AZADO) and 1-Me-AZADO: highly efficient organocatalysts for oxidation of alcohol. *J. Am. Chem. Soc.*, 128 (26): 8412-8413.

[95] Pellissier, H. 2011. Recent developments in dynamic kinetic resolution. *Tetrahedron*, 67 (21): 3769-3802.

[96] Verho, O. and Bäckvall, J. E. 2015. Chemoenzymatic dynamic kinetic resolution: a powerful tool for the preparation of enantiomerically pure alcohols and amines. *J. Am. Chem. Soc.*, 137 (12): 3996-4009.

[97] Rudroff, F., Mihovilovic, M. D., Gröger, H., Snajdrova, R., Iding, H. and Bornscheuer, U. T. 2018. Opportunities and challenges for combining chemo- and biocatalysis. *Nat. Catal.*, 1: 12-22.

[98] Dinh, P. M., Howarth, J. A., Hudnott, A. R. and Williams, J. M. J. 1996. Catalytic racemisation of alcohols: Applications to enzymatic resolution reactions. *Tetrahedron Lett.*, 37 (42): 7623-7626.

[99] Reetz, M. T. and Schimossek, K. 1996. Lipase-catalyzed dynamic kinetic resolution of chiral amines: use of palladium as the racemization catalyst. *Chimia*, 50 (12): 668-669.

[100] Stürmer, R. 1997. Enzymes and transition metal complexes in tandem – a new concept for dynamic kinetic resolution. *Angew. Chem., Int. Ed.*, 36 (11): 1173-1174.

[101] Larsson, A. L. E.; Persson, B. A.; Bäckvall, J. E. 2003. Enzymatic resolution of alcohols coupled with ruthenium-catalyzed racemization of the substrate alcohol. *Angew. Chem., Int. Ed.*, 36 (11). 1211-1212.

[102] Mutti, F. G., Orthaber, A., Schrittwieser, J. H., de Vries, G. J., Pietschnig, R. and Kroutil, W. 2010. Simultaneous iridium catalysed oxidation and enzymatic reduction employing orthogonal reagents. *Chem. Commun.*, 46 (42): 8046-8048.

In: Alcohol Oxidation
Editor: Steffan Paterson
ISBN: 978-1-53614-604-2

Chapter 3

SELECTIVE ALCOHOL OXIDATION BY METAL-ORGANIC FRAMEWORKS AND COVALENT ORGANIC FRAMEWORKS

Sara Abednatanzi, Parviz Gohari Derakhshandeh, Karen Leus and Pascal Van Der Voort
Department of Chemistry, Center for Ordered Materials, Organometallics and Catalysis (COMOC), Ghent University, Ghent, Belgium

ABSTRACT

Metal-Organic Frameworks (MOF, invented in 1999) and Covalent Organic Frameworks (COF, invented in 2005) are two classes of highly porous and designable materials that have been heavily investigated over the last years.

A MOF typically consists of inorganic nodes that are connected by organic linkers to form crystalline, highly porous structures. In this chapter, we will focus on the use of MOFs as heterogeneous oxidation catalysts. MOFs are very promising candidates to replace homogeneous catalysts with sustainable and stable heterogeneous catalysts. The catalytic active function can be either the active metal sites of the MOF itself or can be introduced as an extra functionality in the linker, a dopant or a ship-in-a-bottle complex. As the pore size, pore shape and functionality of MOFs can be designed in numerous ways, shape selectivity and even chiral selectivity can be created.

COFs can be referred to as metal-free porous polymers, except the fact that COFs are (normally) crystalline materials. Many structures have been developed in the last few years, from the original boronate COFs to imine COFs, and CTFs (Covalent Triazine Frameworks) and many others. In the field of oxidation catalysis, the stability is obviously very important. Imine COFs and CTFs are much more stable than MOFs, but their tunability is somewhat lower and the synthesis of many COFs remains a bit cumbersome. However, they open a path to heterogeneous metal-free (photo)catalysis, and that is a very promising route.

1. General Introduction on (Metal-Organic Frameworks) MOFs

Selective catalytic oxidation and asymmetric catalytic oxidations are of industrial importance for fine chemical syntheses. Many homogeneous catalysts, often (transition) metal complexes or oxides are still used in some industrial processes. By varying the metal center and the surrounding ligands of these homogeneous catalysts, the chemo-, regio- and stereo-selectivity can be tuned. Although homogeneous catalysts have the advantage that the reagents can easily approach the catalytically active sites, the separation of the catalyst and products can be cumbersome and expensive. Furthermore, volatile solvents need to be used to recover the catalyst from the reaction mixture. In food and pharmaceutical industry, even the smallest trace of remaining catalyst can render the product unusable. For the design of a heterogeneous variant, several factors need to be taken into consideration. Obviously, the big advantage of a solid catalyst originates in the ease of separation and its recyclability. Nevertheless, for the design of a competitive heterogeneous catalyst, high diffusion dynamics, well-isolated active sites, no leaching and the possibility of a versatile modification of the catalyst should also be taken into consideration.

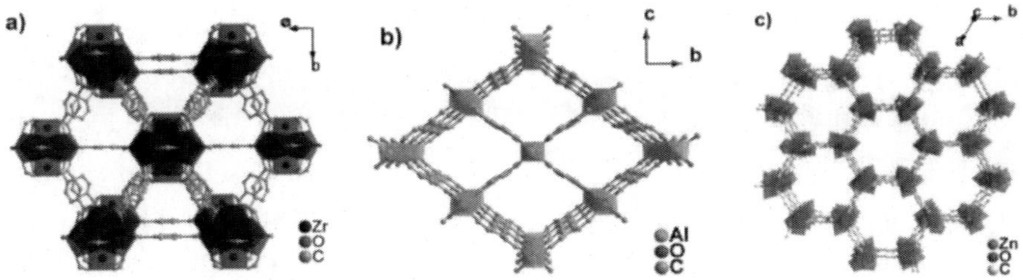

Figure 1. Examples of MOFs with different shapes of pores and channels: (a) UiO-66; (b) MIL-53; and (c) MOF-74. Reprinted from Ref. [3] with permission.

So far, the most often utilized solid catalysts for oxidation reactions are metals, metal oxides or metal complexes immobilized on zeolites, silica, alumina or polymeric resins.

Metal Organic Frameworks (MOFs) are 3D crystalline atomically ordered materials, with an almost unlimited choice of building bricks, allowing a very precise control over morphology, pore size and functionalities in the walls [1, 2]. In Figure 1, some MOFs structures with different shapes of pores and channels are illustrated.

The crystalline nature of MOFs allows a full characterization and therefore a thorough understanding of their behavior is possible. MOFs have typically large surface areas (up to 7000 m^2/g), pore volumes and pore openings up to 9.8 nm [4]. Moreover, MOFs are highly designable due to the chemical versatility and the secondary building blocks that help to build up the framework [5-6]. Due to their attractive properties, many research groups started to explore these materials for several applications. Especially gas storage, gas separation and heterogeneous catalysis are research themes that are quickly emerging [7].

In heterogeneous catalysis, MOFs have been successfully used as catalysts in hydrogenation and oxidation reactions [8], enantioselective reactions [9], photocatalysis [10], and hydrodesulfurization [11]. The highly tunable structure allows the MOF to act both as a heterogeneous catalyst and as a solid support. MOFs with unsaturated metal centers can be directly used as the heterogeneous catalyst. For example, MIL-101 [12] (Fe, Cr, Al) (MIL stands for Materials Institute Lavoisier) and HKUST-1(Cu) [13] (HKUST stands for the Hong Kong University of Science & Technology), which both contain unsaturated metal centers at the nodes of the framework (see Figure 3) have shown good catalytic activity in a wide range of reactions [14-17].

For non-catalytic active MOFs, catalytic sites can be introduced by pre-/post-synthetic modification [18-19]. By using the "ship-in-bottle" approach, catalytically active polyoxometalate clusters [19], metal complexes [20], or nanoparticles [21] can be encapsulated inside the MOF cavities. Furthermore, by pre-/post-synthetic modification on the organic linker, complementary catalytic activity can be introduced [22-23]. With all these efforts, MOF materials can, in principle, combine different catalytic functions in which the unsaturated metal centers can provide Brønsted or Lewis acidity and basicity, or redox active centers. The fine-tunable cavity size (from microporous to mesoporous) permits shape selectivity by avoiding the formation of undesired intermediates or product.

1.1. Designing MOFs for Selective Alcohol Oxidation

MOFs as heterogeneous catalysts offer several possible locations for active sites, including MOFs with inorganic nodes as active sites and MOFs further functionalized by a post-synthetic treatment (Figure 2). Various active sites including metal complexes/oxides, polyoxometalates (POMs) and nanoparticles (NPs) can be introduced into MOFs to design heterogeneous catalysts. Different synthetic routes have been employed to incorporate active sites within the frameworks through immobilization onto the available free sites on the linkers of MOFs or immobilization onto the free metal sites available in the frameworks. Moreover, the active sites can be encapsulated into the framework. These types of active sites have been employed in the use of MOFs as oxidation catalysts, which will be discussed in more details in sections 1.2 and 1.3. Three separate sections (1.4-1.6) will deal with the direct synthesis of modified MOFs, MOF-based composites and photocatalytic oxidation of alcohols by MOFs.

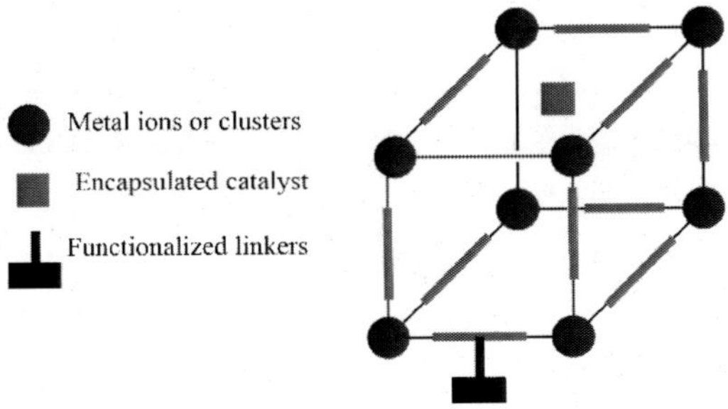

Figure 2. Possible nature and location of active sites in MOFs.

1.2. Synthesis of MOFs Containing Inorganic Nodes as Active Sites

Selective and complete oxidation of alcohols into the corresponding carbonyl compounds is a highly relevant transformation in organic chemistry due to the properties and chemical reactivity of these carbonyl compounds as starting materials in many synthetic methods [24-26]. Traditional processes employ stoichiometric amounts of toxic and expensive inorganic oxidants, mainly iodosylbenzene, sodium hypochlorite and chromium trioxide [27]. Although these strong oxidants improve the yield of the product, their applications are considerably restricted due to high cost, environmental pollution and lower atomic economy [28-31]. Therefore, developing an appropriate catalyst for this transformation is a long-standing research. In the present chapter, the current application of MOFs as active catalysts for the selective oxidation of alcohols is summarized, grouping the reports according to the nature of the active sites present on the MOFs.

As can be seen from Table 1, many examples for heterogeneous oxidation of alcohols with MOFs belong to framework catalysis by the use of coordinatively unsaturated metal (CUM) sites from secondary building units (SBUs). Copper-based MOFs are one of the most studied MOFs for alcohol oxidation. The main advantage of these catalysts is that expensive noble metals are not needed in the process. Cu-BTC (BTC: 1,3,5-benzenetricarboxylate) or HKUST-1 (Hong Kong University of Science and Technology), as a good representative, was first discovered in 1999 [13].

Cu-BTC is a 3-D porous MOF possessing square-shaped channels of 0.9 nm in diameter with water molecules coordinated to the copper centers [32]. Removal of water molecules upon thermal activation creates open Cu^{II} sites capable of attaching to organic molecules (Figure 3).

Cu-BTC owing to its facile synthesis and activation, large specific surface area, relatively large and open pores as well as thermal stability [34] is a good candidate in heterogeneous catalysis. The potential application of Cu-BTC for the aerobic oxidation of alcohols is reported in many studies. One of the first reports has shown the applicability of Cu-BTC for the selective oxidation of a number of benzylic alcohols to the corresponding benzylic aldehydes in moderate to good yields (Table 1, entry 1) [35]. It has been observed that the combination of TEMPO as co-catalyst and Na_2CO_3 as base is necessary to achieve good catalytic activity. In spite of the high catalytic activity of the Cu-BTC/TEMPO system towards benzylic alcohols, the authors have found limitations in terms of scope and catalyst stability. More specifically, the poor reactivity of secondary alcohols was obtained using this catalytic system. The low reactivity of these alcohols is correlated to the inability of substrates/reagents to reach the active sites and catalyst deactivation by poisoning and/or pore blocking. Moreover, powder XRD analysis of the recovered catalyst showed significant changes in the relative intensity of some peaks compared to the fresh catalyst, indicating that Cu-BTC undergoes some structural variations under the experimental conditions.

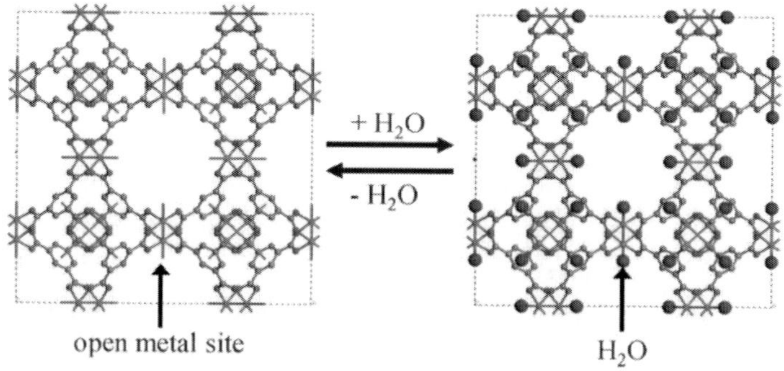

Figure 3. The structure of Cu-BTC MOF. Reprinted from Ref. [33] with permission.

Table 1. Catalytic activities of MOFs containing inorganic nodes as active sites for alcohol oxidation

Entry	Catalytic MOF	Substrate + Conditions	Conv. (%), TON/TOF	Stability evidence	Take home message	Ref.
1	Cu-BTC	Benzylic alcohols, TEMPO, O_2, Na_2CO_3, CH_3CN, 75°C, 22 h	8-89, 1.5	Hot filtration, ICP, XRD, IR	Low reactivity of some alcohols: (i) inability of substrate/reagent to reach the active sites (ii) catalyst deactivation	[35]
2	Mesoporous Cu-BTC	Benzylic alcohols, TEMPO, Na_2CO_3, DMF, O_2, 75°C, 6-22 h	~ 58-99	Reuse, XRD, SEM, BET	Much higher catalytic activity than that of commercial Cu-BTC	[36]
3	Nanosized Cu-BTC	Benzylic alcohols, TEMPO, Na_2CO_3, CH_3CN, O_2, 75°C, 16 h	69-99	Reuse	Enhanced activity: increase of the dispersion introduced by nanosized MOF, size and shape selective	[37]
4	Cu-FMOF	Cyclic alcohols, TEMPO, air, Na_2CO_3, CH_3CN, 75°C, 16 h	26-99	Reuse, hot filtration, IR, XRD, XPS	Only exterior reaction due to the very low surface area	[38]
5	$\{[Cu_3(\mu_3\text{-4-ptz})_4(\mu_2\text{-}N_3)_2(DMF)_2](DMF)_2\}_n$, $\{[Cu(4\text{-ptz})_2(H_2O)_2]\}_n$	Secondary alcohols, TBHP, microwave (MW) irradiation, 80°C, 1 h	38-86, 190-430 (h^{-1})	IR, XRD, elemental analysis	Lower activity using H_2O_2, homogeneous catalysis	[39]
6	FDM-3-7	Benzyl alcohol, TEMPO, N-methylimidazole, CH_3CN, air, 70°C, 12 h	62-99	XRD	Size and shape selective, lost of crystallinities after the catalysis	[40]
7	Cu-MOF-74	Benzylic alcohols, 4-dimethylaminopyridine, TEMPO, O_2, CH_3CN, 70°C, 12 h	53-99	Ruse	Improved broad substrate scope: employment of a homogeneous organic base	[41]

Table 1. (Continued)

Entry	Catalytic MOF	Substrate + Conditions	Conv. (%), TON/TOF	Stability evidence	Take home message	Ref.
8	Cu_2 (bdda) $(OAc)_2 \cdot 1H_2O$	Benzylic alcohols, TBHP, 45°C, 1.5-3 h	75-95	Reuse, hot filtration, TEM, ICP	Nano-sized catalyst: increased dispersion and enhanced catalytic reactivity	[42]
9	$[\{Cu(L1)(DMF)\}.DMF.H_2O]_n$	Benzylic alcohols, TBHP, MW irradiation, 100°C, 0.5 h	~ 13-98, 128-983 (h^{-1})	Reuse, hot filtration, IR, XRD	High catalytic efficiency: The ligand L1 can assist the proton transfer steps	[43]
10	M-BTC (M= Co, Cu, Fe)	Benzyl alcohol, DMF, air, 95°C, 10 h	~ 11-93	XRD, hot filtration	Higher activity of Co-BTC: O_2 activation on Co centers, structural variations after catalysis	[44]
11	$M(bdc)(ted)_{0.5}$ (M= Co, Zn, Ni, Cu)	Benzyl alcohol, DMF, O_2, 90°C, 8 h	~ 6-82	Reuse, IR, XRD, hot filtration	$Co(bdc)(ted)_{0.5}$ showed the highest activity	[45]
12	Co-ZIF-9	Aromatic alcohols, toluene, NaOH, O_2, 150°C, 4 h	~ 35-95	Hot filtration, UV–vis, XRD, ICP	Small pore size: limitation in substrate diffusion	[46]
13	STA-12 (Ni, Co)	Benzylic alcohols, TBHP, ethyl acetate, 60°C, 1.5-3 h	32-100, 9.6-20 h^{-1}	Reuse, IR, XRD	Higher efficiency of STA-12 (Co) than both STA-12 (Ni) and STA-12 (Ni+20% Co)	[47]
14	$([Co_3L2(PTA)_{2.5}(OAc)]\cdot$ solvent)	Benzylic alcohols, TBHP, CH_3CN, 60°C, 16-24 h	77-98	Reuse, XRD	Selective towards aldehyde or ketone depending on the substrate	[48]
15	MIL-100 (Sc, Fe)	Secondary alcohols, TBHP, toluene/decene, room Temp., 6 h	6-99	Leaching	Higher catalytic activity of mixed-metal MIL-100 (Sc, Fe) compared to MIL-100 (Sc)	[49]

Entry	Catalytic MOF	Substrate + Conditions	Conv. (%), TON/TOF	Stability evidence	Take home message	Ref.
16	MIL-100 (Fe)-NH_4F	5-Hydroxymethylfurfural, TEMPO, CH_3CN, O_2, NaN_2, 75°C, 6 h	100	Reuse, leaching, hot filtration, IR, XRD	Higher catalytic activity compared to MIL-100 (Fe) due to the higher surface area, loss of activity in the 2nd run	[50]
17	V-NDC	Benzyl alcohol, TBHP, toluene, 90°C, 4 h	73	Reuse	From oil refinery waste into active V-based catalysts	[51]
18	$Pd(2\text{-}PYMO)_2$	Cinnamyl alcohol, toluene, 90°C, 20 h	99	-	Atmospheric air as the sole oxidant	[52]
19	Ce-UiO-66-BDC	Benzyl alcohol, TEMPO, O_2, CH_3CN, 110°C, 24 h	88	Hot filtration, XRD, ICP	Synergy with TEMPO, activation at high temperature	[53]
20	Ce-MOF-808	Benzyl alcohol, TEMPO, CH_3CN, O_2, 110°C, 7 h	97	XRD, hot filtration	Higher catalytic activity than Ce-UiO-66: large pore size and low connectivity of Ce-MOF-808	[54]

BTC: 1,3,5-benzenetricarboxylate, TEMPO: 2,2,6,6-tetramethylpiperidine-N-oxyl, Cu-FMOF: $\{[Cu_2(1,2\text{-}BDC)_2(Fbtx)_2]\cdot 3H_2O\}_n$, Fbtx: 1,4-bis(1,2,4-triazole-1-ylmethyl)-2,3,5,6- tetrafluorobenenze), BDC: 1,4-benzenedicarboxylic acid, 4-ptz: 5-(4-pyridyl)tetrazolate, FDM-3-7: trinuclear copper pyrazolate building blocks, MOF-74: [M_2(DOT): DOT = dioxidoterephthalate], H_2bdda: 4,4′-[benzene-1,4-diylbis (methylylidenenitrilo)] dibenzoic acid, L1: 5-{(pyridin-4-ylmethyl)amino} isophthalic acid, ted: triethylenediamine, ZIF-9: zeolitic imidazolate framework, STA‐12 (Co): ($Co_2L.7.4H_2O$), STA‐12 (Ni): ($Ni_2L.6.6H_2O$) and L: N,N'-piperazine bis(methylphosphonic) acid, L2: bent imidazole-bridged ligand, PTA: *p*-phthalic acid, MIL: Materials Institute Lavoisier, NDC: 1,4-naphthalenedicarboxylate, 2-PYMO: 2-hydroxypyrimidinolate, UiO-66: (University of Oslo) Zr-based MOF, Ce-MOF-808: $[Ce_6O_4(OH)_4(btc)_2(OH)_6]$.

One of the main advantages of MOFs is the ability to control and modify the pore size and surface area to find improved properties in catalysis. Within this regard, Peng et al. Could successfully synthesize the mesocellular Cu-BTC with large mesopores (13–23 nm) using CO_2-expanded liquids as switchable solvents (Figure 4) [36]. The porosity properties of this MOF could be easily modulated by CO_2 pressure. Larger mesopores and thinner pore walls were obtained at a higher CO_2 pressure which further resulted in the higher activity of the catalyst. By using the mesoporous Cu-BTC synthesized at high CO_2 pressure (6.6 Mpa), benzyl alcohol converted completely to benzaldehyde after 3 h. The activity of this catalyst was higher than that synthesized in pure DMF and much higher than the commercial Cu-BTC (10% conversion) under the same reaction conditions.

It has been proved that the size and shape of crystals are important features in liquid-phase catalysis. For instance, the diffusion length of nanoscale MOFs as catalysts is decreased in comparison to its bulk analogs. By applying specific synthesis processes, Cu-BTC with different dimensions from nm to μm can be achieved with different morphologies (spherical or octahedral) [37]. Nanosized Cu-BTC was further utilized as a heterogeneous catalyst allowing the efficient oxidation of alcohols (99% yield for benzyl alcohol). However, the aerobic oxidation of secondary alcohols did not proceed under the optimal reaction conditions. Interesting size and shape selectivity was studied. The increase of the size of aromatic rings decreases yields of the corresponding aldehydes. In the case of benzyl alcohol having a size of 4.3 Å that is smaller than the pore size, the yield was up to 99%. By increasing the number of the aromatic rings (bigger than the available pore dimensions of Cu-BTC) much lower yields were achieved (33% yield using 9-Anthracenemethanol).

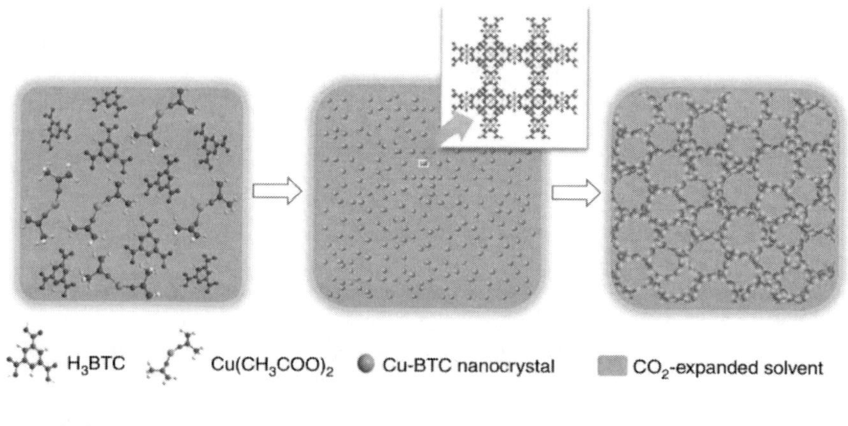

Figure 4. Schematic illustration for mesocellular Cu-BTC formation. Reprinted from Ref. [36] with permission.

Apart from the well-known Cu-BTC MOF, many attempts have been devoted to the synthesis of new copper-based MOFs to be applied in the alcohol oxidation catalysis. A new copper-based MOF with mixed linkers $\{[Cu_2(1,2\text{-}BDC)_2(Fbtx)_2]\cdot 3H_2O\}_n$ (denoted as Cu-FMOF) was synthesized by a hydrothermal method and its catalytic activity was evaluated in alcohol oxidation (Table 1, entry 4) [38]. Although various benzylic alcohols with electron-donating or -withdrawing functionalities gave the corresponding aldehydes in good to excellent yields, efforts to oxidize secondary alcohols were not successful. All of these observations clearly demonstrate that the aerobic oxidation of secondary alcohols to their ketones using copper-based MOFs still remains a challenge. As can be concluded from Table 1, the presence of TEMPO as co-catalyst for the aerobic oxidation of alcohols is vital to achieve high catalytic activity. Chen et al. reported a proposed reaction mechanism for the TEMPO-assisted oxidation of alcohol catalyzed by Cu-FMOF [38]. According to this mechanism benzyl alcohol reacts with the five-coordinated Cu(II) ions and this intermediate reacts with the TEMPO radical. Subsequently, deprotonation of the alcohol generates aldehyde and Cu(I)-FMOF. The spontaneous oxidation of TEMPOH to TEMPO under oxygen helps further to regenerate Cu(II)-FMOF.

$\{[Cu_3(\mu_3\text{-}4\text{-}ptz)_4(\mu_2\text{-}N_3)_2(DMF)_2](DMF)_2\}_n$ and $\{[Cu(4\text{-}ptz)_2(H_2O)_2]\}_n$ with 3D and 2D coordination networks, are two other new MOFs [39]. Figure 5 shows the schematic view of synthesis, structure formulae and topological representations of these frameworks. In contrast to the previous reports, the catalytic investigations revealed that these MOFs are active catalyst precursors towards the oxidation of various secondary alcohols with *tert*-butyl hydroperoxide, resulting in the yields of the corresponding ketones up to 86% (TOF = 430 h^{-1}) and 58% (TOF = 290 h^{-1}) in the oxidation of 1-phenylethanol and cyclohexanol, respectively, and in the absence of any added solvent or additive. Unfortunately, powder XRD analysis of the recycled catalyst confirmed that both catalysts lost their crystalline nature compared with the fresh ones and therefore behave as precursors of homogeneous catalytically active species.

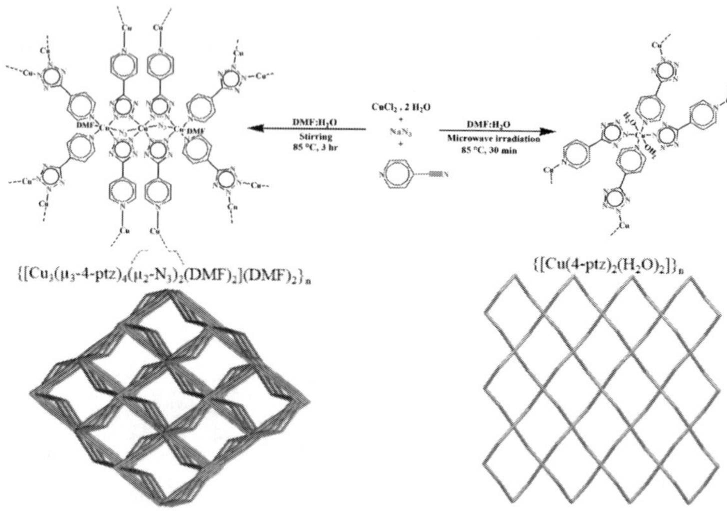

Figure 5. Synthesis and structural formulae of 3D and 2D Cu-based compounds and their Topological representations. Reprinted from Ref. [39] with permission.

MOFs with Co ions in the nodes have shown high activity to proceed oxidation of alcohols. In general, Co-based MOFs show higher efficiency in the aerobic oxidation of alcohols compared to Cu-based MOFs. As can be seen from Table 1, for Cu-based MOFs the presence of TEMPO as co-catalyst is vital to obtain a good catalytic activity using oxygen as oxidant. Switching to Co-based MOFs, a good catalytic activity can be achieved without the use of any co-catalyst. This behavior is well studied by using a series of M-BTC frameworks with cobalt, copper and iron as the active metal ions (Table 1, entry 10) [44]. Under the optimal reaction conditions, the conversion of benzyl alcohol can reach to 92.9% using Co-BTC with a selectivity of 97% towards benzaldehyde. Compared with Co-BTC, Cu-BTC and Fe-BTC show much lower catalytic activity resulting in only 24 and 11% conversions, respectively. Although Co-BTC has a lower surface area compared to both Cu-BTC and Fe-BTC, the results confirm that the catalytic behavior is not related to the BET surface area. The results show that the oxygen species on the surface of the catalyst are involved in the reaction process, and the catalyst can effectively activate the oxygen molecules in the air to realize the catalytic conversion of alcohols. It is found that Co-BTC can effectively activate the oxygen molecules resulting in the higher conversion of alcohol compared to Cu and Fe metal ions. Co-BTC suffers from the low stability under the reaction conditions as powder XRD showed a major change in crystallinity after two consecutive runs compared to the fresh catalyst. Moreover, the catalytic activity is reduced to 88% after the third cycle.

It is well documented that Co-based compounds are highly efficient oxidation catalysts. A similar comparative study has shown the difference in catalytic activity of a series of $M(BDC)(TED)_{0.5}$ MOFs (M = Co, Zn, Ni, Cu) (Figure 6) for the benzyl alcohol oxidation [45]. The order of increasing reactivities of the catalyst based on their metal ions are as Co > Ni > Cu > Zn with conversions of 81.8, 25.2, 8.8 and 5.6% respectively (Table 1, entry 11). The reaction is done in DMF at 90°C using oxygen as the sole oxidant. As mentioned earlier, these Co-based compounds do not require either TEMPO or base to promote the reaction under the applied conditions. As can be concluded from these reports the activity of the metal ions of the frameworks has a direct impact on the catalytic efficiency of catalysts. However, the activity of this catalyst is examined only for benzyl alcohol and no studies were done to expand the substrate scope to other more challenging substrates such as secondary alcohols. The recycled $Co(BDC)(TED)_{0.5}$ catalyst could be successfully reused for three runs without obvious loss of the catalytic activity and selectivity. Furthermore, no changes are observed for the crystallinity between the fresh and three times recycled, proving the high stability of the catalyst.

As it can be seen here, MOFs are active catalysts for the selective oxidation of alcohols. However, there is still room to design more efficient catalysts concerning both the activity and stability. In principle, one of the interesting properties of MOFs is the possibility to introduce more than one metal in the framework, labeled as mixed-metal MOFs. These mixed-metal MOFs are tunable in a very systematic way to obtain the desired properties in flexibility, catalytic activity, sensing efficiency, etc. Within this context, MIL-100 (Sc, Fe) as a three dimensional M (III) trimesate catalyst, in which Fe^{3+} cations as well as Sc^{3+}, occupy framework sites has been synthesized [49]. The structure of MIL-100 is shown in Figure 7. The metal sites as CUM can effectively be used in oxidation catalysis. The MIL-100 framework is a great potential host for the active sites due to its large surface area with accessible cages. The catalytic activity of this catalyst was tested for the oxidation of 2,2,2-trifluoro-1-(2-methyl-1H-indol-3-yl)ethanol and related compounds (Table 1, entry 15). In contrast to the poor performances of MIL-100 (Sc) and Cu-BTC (8 and 9% yield respectively), the mixed MIL-100 (Sc, Fe) material gave 80% yield for the corresponding ketone even at Fe contents as low as 40 mol %.

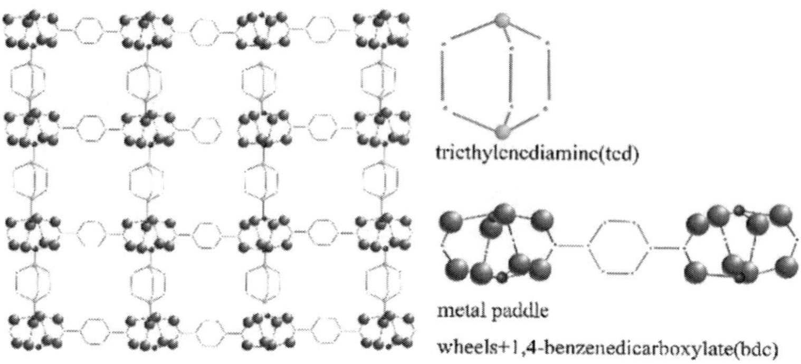

Figure 6. Schematic presentation for the construction of typical $M(BDC)(TED)_{0.5}$ MOFs (M = Co, Zn, Ni, Cu), M (red), O (green), N (yellow), and C (gray). Reprinted from Ref. [45] with permission.

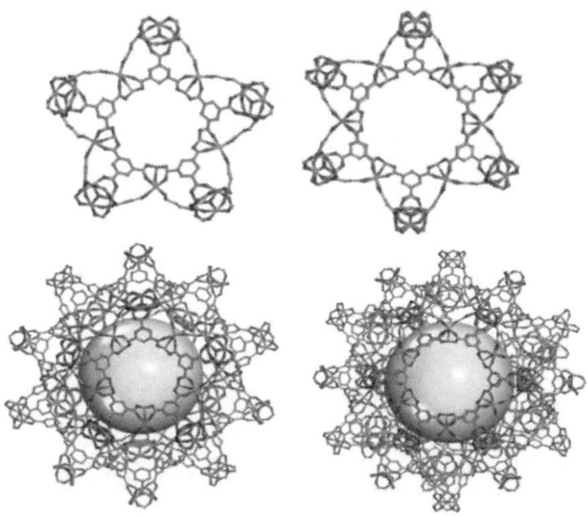

Figure 7. Structure of MIL-100, showing the assembly of trimer $Sc_3O(O_2C\text{-})_6$ units via supertetrahedra to give networks containing two kinds of cage. Reprinted from Ref. [55] with permission.

1.3. Synthesis of MOFs Further Functionalized by a Post-Synthetic Treatment

In section 1.2 we showed the applicability of MOFs, which are synthesized by direct approaches without a further modification, as heterogeneous catalysts for alcohol oxidation. The direct synthesis of MOFs with active sites within the framework is still limited since a suitable attention is required to linker design and node composition. Moreover, a number of reported MOFs suffer from the low chemical stability under the reaction conditions. On the other hand, the activity of these catalysts is mostly limited to benzylic alcohols. Very few studies could successfully increase the activity of MOFs by modulating internal surface areas and cavity/pore sizes or control over the size of particles.

Another route for obtaining MOFs containing active sites for catalysis is by post-synthetic modification (PSM). Rather than the direct synthesis of MOFs using functionalized ligands, the chemical modification of frameworks can be done after their synthesis. Since the time that Wang and Cohen performed the first detailed study of PSM in 2007 [56], considerable efforts have been devoted to the development of efficient heterogeneous catalysts. The main advantage of the PSM route compared to the direct synthesis is the excellent control over the kinds and number of functional groups that can be incorporated into the framework. A large variety of metals can be incorporated into the MOF pores via three main strategies including (a) covalent modification (chemical modification of the organic linker of the MOF), (b) coordinate covalent modification (ligands coordination to CUMs or metals coordination to the organic linker of MOFs), (c) covalent and coordinate covalent modification (the organic linker within the MOF can be appended with a ligand and consequently metallated) [57] (Figure 8). In this section, we will describe some of the efforts towards the incorporation of metal complexes/oxides, polyoxometalates, porphyrins and nanoparticles resulting in heterogeneous catalysts for alcohol oxidation.

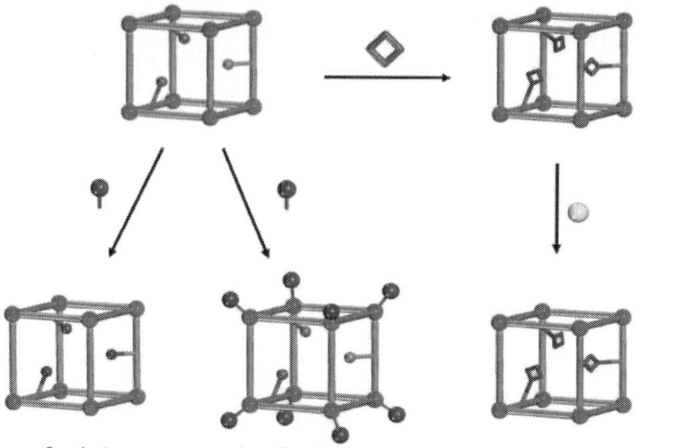

Figure 8. Representation of different PSM routs of MOFs. Reprinted from Ref. [57] with permission.

1.3.1. Immobilization of Metal Complexes/Oxides into MOFs

As can be seen from Table 2, copper is considered to be one of the most studied catalysts for use in the oxidation of alcohols. With this regard, different MOFs have been applied as support for copper complexes including UiO-66, MOF-NH_2 and PCN-700. Among the well-known MOFs, Zr-based frameworks have been broadly examined due to their excellent chemical stability in aqueous solutions, as well as basic or acidic media. As an example, UiO-66 (University of Oslo) is one of the widely applied MOFs in catalysis prepared by the reaction of zirconium salts with BDC linkers. UiO-66 is amongst the resistance frameworks having high chemical and thermal stability. UiO-66-NH_2 was functionalized with salicylaldehyde using a post-synthetic strategy to attach various Cu(II) salts and their performance was tested in the oxidation of alcohols [58]. PCN-700 is another highly stable Zr-based framework with eight-connected $Zr_6O_4(OH)_8(H_2O)_4$ clusters based on 2,2'-dimethylbiphenyl-4,4'-dicarboxylate as the linker [59]. This MOF is an excellent example of engineering the pore environment using different functionalities which will be discussed later. The presence of the open bipyridine moieties provides suitable sites that readily react with Cu(I) to form the metalated framework

[59]. MOF-NH_2 which is a Zn-based framework was synthesized through the reduction of an azide functionalized MOF (MOF-N_3). The incorporation of $CuCl_2$ to the amine functionalized MOF gave the Cu(II)/MOF-NH_2 material [60]. One of the main attractive characteristics of these MOFs is their high surface areas of 1338, 2130 and 582 m^2/g respectively, as well as their tunable pore size and topology. These features make them good candidates to introduce desired functionalities. After the introduction of Cu(II) or Cu(I) complexes through PSM, all of the catalysts present high catalytic activity for the aerobic oxidation of alcohols resulting in a full conversion and selectivity of benzyl alcohol to benzaldehyde (Table 2, entries 1-3). In general, the presence of TEMPO is necessary in combination with Cu complexes to have a good catalytic activity for these catalysts which is in agreement with the studies on copper-based MOFs synthesized through the direct synthesis [35-38]. A schematic illustration of the structure of the heterogeneous copper-based catalysts is presented in Figure 9. Different from some of the copper-based MOFs described in section 1.2 showing low stability, all of these studies confirmed that the structure of the modified MOFs is well preserved after the catalytic reaction. Moreover, Cu(II)/MOF-NH_2 catalyst exhibited a very good performance for a broad range of substrates including inert alcohols [60]. It is worth noting that Cu(II)/MOF-NH_2 could act as an efficient catalyst even without the presence of a base. In this catalytic system, the reason for the high efficiency might be due to the fact that the MOF containing an amino group had adequate basicity for deprotonation, and the oxidation could occur without any external base.

Besides the synthesis of MOFs with high catalytic activity, it is highly desired to apply the unique properties of MOFs and more specifically their PSM to design materials for more specialized applications. This can be achieved by decorating the pore environment of MOFs with multiple functional groups. An elegant example is PCN-700 which we mentioned its catalytic activity in alcohol oxidation. Using PCN-700 with coordinatively unsaturated Zr_6 clusters, 12 linkers with different lengths and functionalities were installed in the parent framework [59] (Figure 10). The length of the linkers has a direct impact on the MOF porosity and channel sizes. This leads

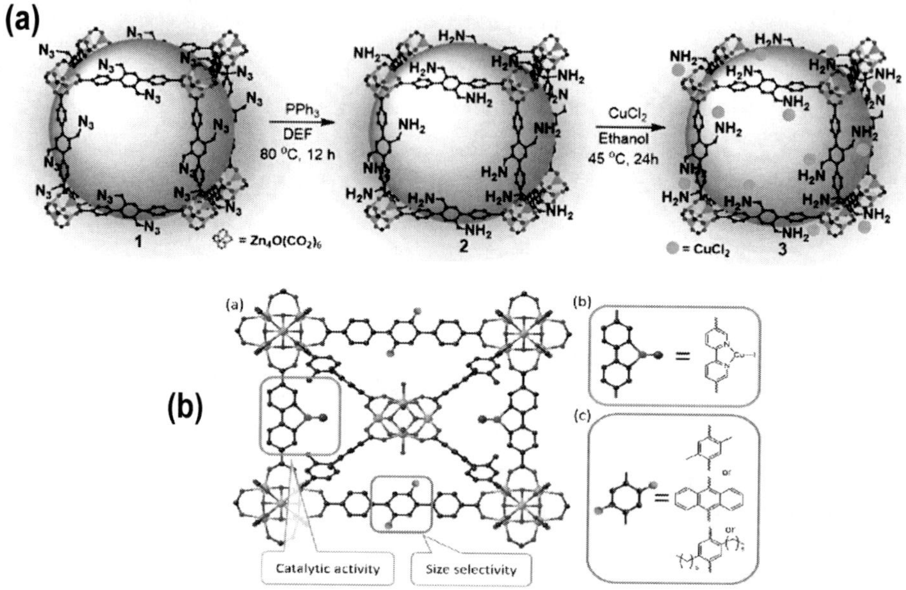

Figure 9. Schematic illustration of modified MOFs with copper complexes: (a) Cu(II)/MOF-NH_2, (b) PCN-700-BPYDC(Cu)-TPDC-R_2. Reprinted from Ref. [59-60] with permission.

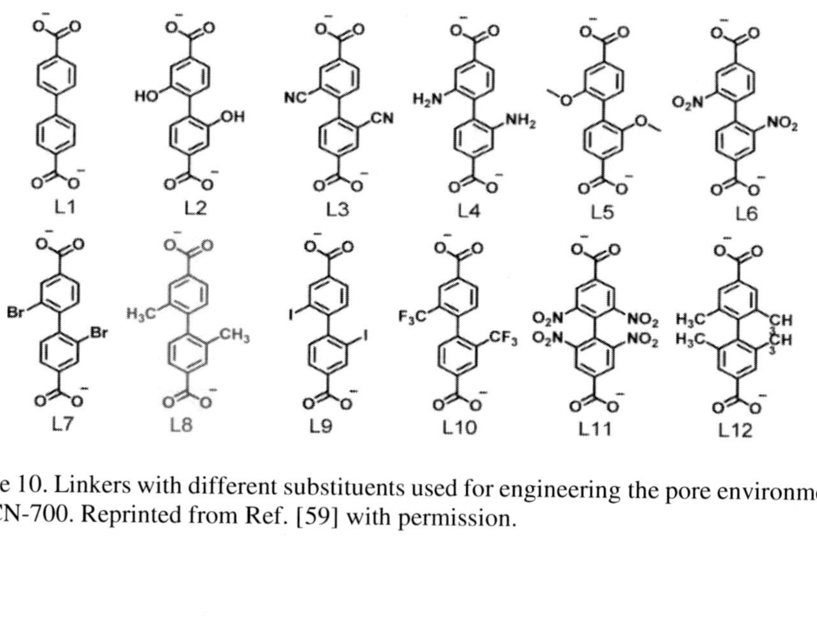

Figure 10. Linkers with different substituents used for engineering the pore environment of PCN-700. Reprinted from Ref. [59] with permission.

Table 2. Catalytic activities of MOFs containing immobilized metal complexes/oxides for alcohol oxidation

Entry	Catalytic MOF	Substrate + Conditions	Conv. (%), TON/TOF	Stability evidence	Take home message	Ref.
1	UiO-66-sal-$CuCl_2$	Benzylic alcohols, TEMPO, $NaHCO_3$, CH_3CN, O_2, 60°C, 24 h	45-99	Reuse, hot filtration, SEM, IR, TGA	Higher activity compared to UiO-66-sal-$Cu(NO_3)_2$ and UiO-66-sal-$Cu(OAc)_2$	[58]
2	PCN-700-BPYDC(Cu) & PCN-700-BPYDC(Cu)-TPDC-R_2 (R= Me, Ph, Hex)	Aromatic alcohols, TEMPO, N-methylimidazole, CH_3CN, air, room Temp., 6 h	1-99	Reuse	Pore environment engineering resulted in a size selective catalyst	[59]
3	Cu(II)/MOF-NH_2	Alcohols, TEMPO, CH_3CN, air, 70°C, 5.5-11.5 h	48-98	Hot filtration, reuse, ICP, IR, XRD, XPS	Broad substrate scope: big pore size and surface functionality, base-free catalysis	[60]
4	$CoAl(OH)_2$@NU-1000	Benzyl alcohol, TBHP, CD_3CN, 80°C, 2 h	~ 52, 220	Reuse, SEM, TEM, N_2 adsorption	Higher efficiency compared to their homometallic counterparts	[61]
5	$Ir^{III}Cp^*Cl$@COMOC-4	Benzyl alcohol, Cs_2CO_3, iodobenzene, toluene, 150°C, 24 h	54	Reuse, hot filtration, ICP, XRD	Effect of the synthetic way on the surface area and Iridium loading (pre-functionalization *vs* post-modification)	[62]
6	$RuCl_3$@MIL-101(Cr)–tpy	Benzylic alcohols, H_2O_2, H_2O, 100°C, 5.5-9 h	95-99	Hot filtration, AAS, reuse	Higher activity compared to its homogeneous counterpart	[63]

Table 2. (Continued)

Entry	Catalytic MOF	Substrate + Conditions	Conv. (%), TON/TOF	Stability evidence	Take home message	Ref.
7	CuI/MIL-101-N-2-pyc	Primary and secondary alcohols, TEMPO, N-methylimidazole, CH_3CN, air, room Temp., 1-12 h	66-99	Reuse, leaching, XRD,	Broad substrate scope: ligand optimization to achieve high efficiency	[64]
8	FePcS/MIL-101	Benzyl alcohol, TBHP, CH_3CN, 60°C, 2 h	32	Hot filtration, elemental analysis, IR, Raman	The oxidation rate decreased with increasing FePcS content	[65]
9	Fe/MIL-101	Benzylic, primary and secondary alcohols, H_2O_2, H_2O, 15 min, 300 W, 110-120°C	10-99	Reuse, hot filtration, XRD, N_2 physisorption, ICP	Enhanced Lewis acidity upon Fe incorporation	[66]
10	Pt/MOF-5	Piperonyl alcohol, NaOH, $(Pb(OAc)_2)$, air, 80°C, 1 h	96	-	The activity and selectivity was comparable with Pt/C (C: activated carbons)	[67]
11	MOF-253-Ru	Secondary alcohols, $PhI(OAc)_2$, CH_2Cl_2, room Temp., 40°C, 2-5 h	64-99, 16-99 (h^{-1})	Hot filtration, ICP, reuse, XRD, BET	High efficiency with samples containing low Ru loadings and higher surface areas	[68]
12	V-Nu-1000	4-methoxybenzyl alcohol, O_2, toluene, Et_3N, 105°C, 8 h	56, 2.6 (h^{-1})	Leaching, ICP, reuse, XPS, SEM, EDS, BET, XRD	Higher activity compared to high surface area $V\text{-}ZrO_2$	[69]
13	UiO-67-$FeCl_3$	Cyclic alcohols, TEMPO, $NaNO_2$, O_2, CH_3CN, room Temp., 12 h	58-99	Reuse, hot filtration, ICP, XRD, IR	Pyridine moiety of the framework as an organic base: facilitating the conversion of benzyl alcohol	[70]
14	UiO-66-CrCAT	Secondary alcohols, TBHP, neat or C_6H_5Cl, 70°C, 8-24 h	70-99, 2.94-34.02 (h^{-1})	Hot filtration, ICP, reuse, XRD, SEM, XANES	Broad scope of substrates	[71]

Entry	Catalytic MOF	Substrate + Conditions	Conv. (%), TON/TOF	Stability evidence	Take home message	Ref.
15	Cu(I)/UiO-66-NH-PC	Primary, secondary and benzylic alcohols, TEMPO, N-methylimidazole, CH_3CN, air, room Temp., 1-9 h	68-99	Reuse, hot filtration, XRD, SEM	Broad substrate scope, high yield for the transformation of secondary alcohols to ketones under mild conditions	[72]
16	PW/DAIL/MIL-101(Cr)	Benzylic alcohols, TBHP, $CHCl_3$, 100°C, 4-6 h	45-100, 900-2000	Hot filtration, reuse, ICP, XRD, DRIFT	High catalytic efficiency: the ability of the DAIL groups to form hydrogen bonds and enhancing the accessibility of TBHP	[73]
17	PW_4/DAIL/MIL-100(Fe)	Benzyl alcohol, TBHP, $CHCl_3$, 100°C, 6 h	92, 613	Hot filtration, reuse, ICP, XRD	High catalytic efficiency: the ability of the DAIL groups to form hydrogen bonds and enhancing the accessibility of TBHP	[74]

PCN: Porous coordination network, Zr-based MOF, BPYDC = 2,2′-bipyridne-5,5′-dicarboxylate, TPDC = terphenyl-4,4″-dicarboxylate, NU-1000: a Zr-based MOF consisting of $Zr_6(\mu_3\text{-}O)_4(\mu_3\text{-}OH)_4(H_2O)_4(OH)_4$ nodes and 1,3,6,8-(*p*-benzoate)pyrene (TBAPy) linkers, COMOC-4: nanoporous Ga-based MOF, Ga(OH)(BPYDC), Cp*: 1,2,3,4,5-pentamethylcyclopentadienyl, tpy: 2,2':6',2"-terpyridine, pyc: pyridinecarbaldehyde, FePcS: Iron tetrasulfophthalocyanine, MOF-5: $[Zn_4O(BDC)_3]$, MOF-253: Al^{3+}-based MOF with open bipyridine sites $[Al(bpydc)(OH)\cdot 0.5H_2O]$, DAIL: dual amino-functionalized ionic liquid.

to a highly tunable multifunctional MOF system that is capable of size selective catalysis. By installation of TPDC with different functional groups (methyl, phenyl, and hexyl) within the PCN-700-BPYDC-TPDC-R_2 catalyst (BPYDC = 2,2′-bipyridine-5,5′-dicarboxylate, TPDC = terphenyl-4,4″-dicarboxylate), the size selectivity can be fine-tuned. The TPDC partially blocks the channel limiting the diffusion of substrates as well as the accessibility of the Cu^+ as the active center. A reduction in yields of alcohols was observed by increasing the size of substituents on the TPDC linker.

Using the catalyst with TPDC-Hex_2, the 9-anthracene alcohol is almost excluded by the small MOF cavity, which explains the lowest yield (< 1%). This comprehensive study proves the importance of the linker installation method to decorate the pore environment of MOFs resulting in catalysts with different activity and selectivity.

From these studies, it can be concluded that a high surface area and the presence of accessible pores are required for the PSM of MOFs as it has an influence on catalytic activity and selectivity. Recently, Abednatanzi et al. reported the heterogenization of an Ir^{III} complex by the modification of a nanoporous gallium 2,2'-bipyridine-5,5'-dicarboxylate (H_2bpydc) MOF, denoted as COMOC-4 [62]. The introduction of the Ir complex was performed successfully through two different strategies (post- and pre-functionalization) as shown in Figure 11. It was found that the applied functionalization route for the introduction of metal complexes has a direct influence on the properties of the prepared MOFs. Higher surface area and Ir loading were obtained using the pre-functionalization method in which the modified bipy linker with Ir complexes are directly used to synthesize COMOC-4 framework. The obtained Ir^{III}Cp*Cl@COMOC-4 catalyst showed good catalytic activity (54% conversion) for the selective aerobic oxidation of benzyl alcohol to benzaldehyde. However, the presence of iodobenzene as co-oxidant was vital to obtain a good catalytic activity.

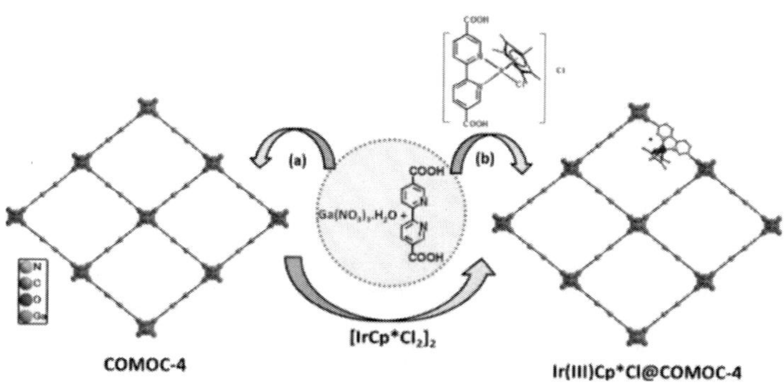

Figure 11. Schematic illustration for the preparation of Ir^{III}Cp*Cl@COMOC-4 by a) post-functionalization and b) pre-functionalization. Reprinted from Ref. [62] with permission.

1.3.2. Immobilization of Polyoxometalates (POMs) into MOFs

Polyoxometalates (POMs) are an infinite class of anionic transition metal oxide clusters and have been widely employed as intrinsic oxidation catalysts owing to their tenability of redox and acid properties [75-76]. MOFs with high surface areas and large pores are good candidates for heterogenization of POMs. MIL-101(Cr), a three-dimensional chromium(III) terephthalate MOF, possesses regular mesoporous cages of 29 and 34 Å, which are accessible through windows of ca. 12 and 16 Å [77] (Figure 12). Interestingly, the existence of coordinatively unsaturated chromium sites (CUSs) in the framework provides opportunities for the incorporation of active species.

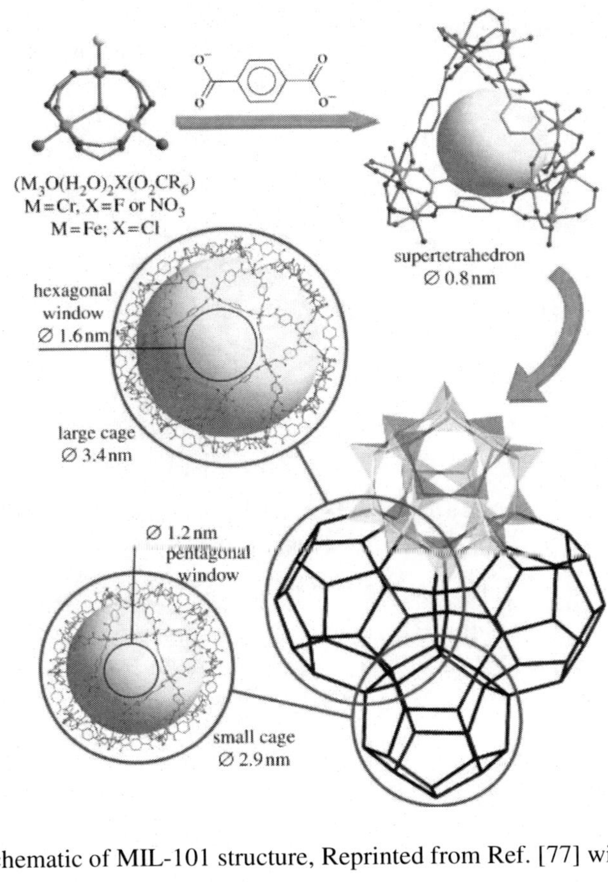

Figure 12. Schematic of MIL-101 structure, Reprinted from Ref. [77] with permission.

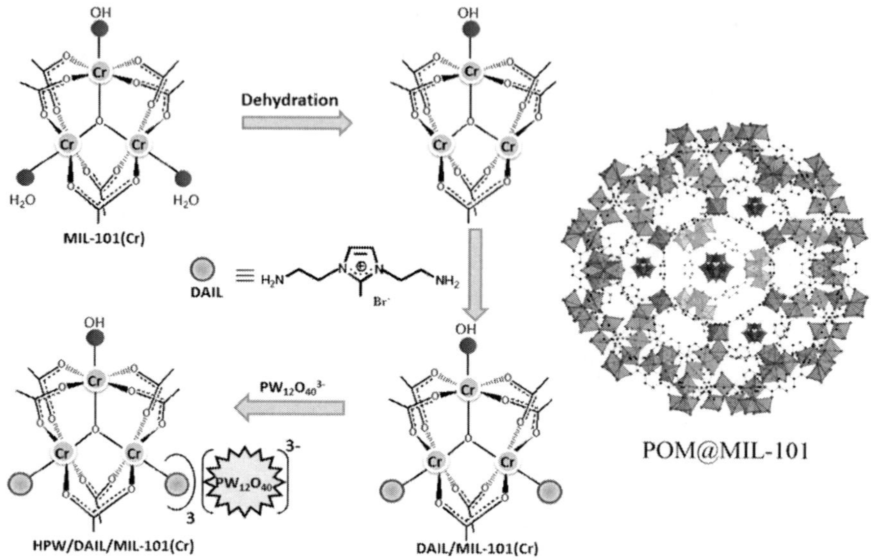

Figure 13. Schematic illustration of the preparation of PW/DAIL/MIL-101(Cr). Reprinted from Ref. [73] with permission.

Recently, phosphotungstic acid ($H_3PW_{12}O_{40}$, HPW) was immobilized into the nanocages of a dual amino-functionalized ionic liquid (DAIL)-modified MIL-101(Cr) framework under mild conditions [73]. The PW cluster, with dimensions of about 12 Å, has the appropriate size to diffuse through the larger windows of the framework. The obtained PW/DAIL/MIL-101(Cr) was evaluated as a catalyst in the oxidation of benzyl alcohol using TBHP in chloroform. In comparison with the HPW/MIL-101(Cr) material, PW/DAIL/MIL-101(Cr) showed an enhanced catalytic performance (95% conversion *vs* 70%) in the selective oxidation of benzyl alcohol with complete selectivity towards benzaldehyde. This was mainly attributed to the ability of the DAIL groups to form hydrogen bonds, thus enhancing the accessibility of TBHP. A schematic illustration for the preparation of the catalyst is outlined in Figure 13.

1.3.3. Nanoparticles (NPs) Embedded in MOFs

Apart from the use of transition metal-based catalysts, supported metal NPs are well recognized for the selective oxidation of alcohols using oxygen in particular (or air, even better), as the cheapest and less polluting stoichiometric oxidant. The embedding of NPs within suitable supports is of great significance to avoid their aggregation, followed by deactivation. MOFs form an exciting new scaffold to anchor NPs offer many advantages in comparison to the traditional support materials. MOFs offer good accessibility of the NPs during catalysis due to their large surface areas, high porosities and pore volumes. In addition, the chelating groups in functionalized MOFs can be used to anchor the NPs inside the pores and prevent agglomeration of NPs, thus controlling their size, shape and size distribution. These features allow an easy selection of the suitable MOF material depending on the size of the NPs required. Several NPs such as Au, Pd, Pt, and Mn_3O_4 have so far been incorporated into MOFs using several synthesis techniques (Table 3).

As can be seen from Table 3, Au NPs are widely applied for the aerobic oxidation of alcohols and they showed high catalytic activity and selectivity towards aldehydes even under base-free conditions [78-80] (Table 3, entries 1-3). As an example, the amino-functionalized UiO-66 was applied for a good distribution of $HAuCl_4$ as the gold precursor over the outer and inner surfaces of the framework [78]. The great distribution of $HAuCl_4$ guarantees that the gold nanoparticles disperse well over the surface of the nano-MOFs (Figure 14). The Au@UiO-66 (NH_2) catalyst showed high catalytic activity (yield= 74-94%) and selectivity for benzylic alcohols without the addition of any base. As a control experiment, the catalytic performance of unsupported gold nanoparticles and Au@TiO_2 catalysts were evaluated. The obtained results confirmed that these catalysts did not function well in the absence of the base under the optimized reaction conditions. The reason for the better catalytic activity of Au@UiO-66 (NH_2) was attributed to the great dispersion of gold nanoparticles in a large surface area MOF support and the narrow distribution of gold nanoparticles without self-aggregation. Moreover, the authors proposed that the amino groups on the MOF act as a weakly basic promoter, which allows for the efficient aerobic oxidation without the addition of a basic co-catalyst.

Besides the use of Au NPs, the use of Pd NPs supported on MOFs has been widely studied for liquid phase oxidation reactions (Table 3, entries 9-14). An interesting example of the alcohol oxidation by Pd NPs was reported

Table 3. Catalytic activities of MOFs containing NPs for alcohol oxidation

Entry	Catalytic MOF (NPs size)	Substrate + Conditions	Conv. (%), TON/TOF	Stability evidence	Take home message	Ref.
1	Au@UiO-66 (NH_2) (~ 3 nm)	Benzylic alcohols, O_2, DMF, 100°C, 1 h	74-99	Reuse, ICP, XRD, IR, TEM, SEM	Higher catalytic activity compared to Au NPs and Au@TiO_2, base-free catalysis	[78]
2	Au@UiO-66-$NaBH_4$ (7 nm)	Benzyl alcohol, O_2, toluene, 100°C, 9.5 h	94	Reuse, XRD, N_2 sorption, XPS, TEM	An efficient catalyst under O_2 and base-free conditions	[79]
3	Au/MIL-101 (2.3 ± 1.1 nm)	Benzylic, allylic and aliphatic alcohols, O_2, toluene, 80°C, 1-35 h	23-99	Reuse, TEM, AAS, leaching	Broad substrate scope, base-free catalysis, highly active and selective	[80]
4	Au/UiO-66 (5 nm)	Benzyl alcohol, O_2, K_2CO_3, water, 80°C, 10 h	56.3, 955 h^{-1}	Reuse, TG, TEM	Promotion effect of water: improving the reaction efficiency	[81]
5	Au_{11}@ZIF-8 (150)	Benzyl alcohol, TBHP, K_2CO_3, toluene, 80°C, 24 h	83.8	Reuse	High catalytic performance: highly dispersed ultrasmall Au NPs, least steric hindrance	[82]
6	Mn_3O_4/MIL-101 (3 nm)	Benzyl alcohol, O_2, toluene, 100°C, 4 h	38.7, 0.84 h^{-1}	Reuse, XRD, IR, ICP	Higher catalytic activity compared to Mn_3O_4 due to its smaller size (3 nm *vs* 12 nm)	[83]
7	Pt@MOF-177 (2-5 nm)	Benzylic alcohols, air, room Temp., 24 h	50-99, 351-968	Reuse, XRD	Recyclability was unsatisfactory	[84]

Entry	Catalytic MOF (NPs size)	Substrate + Conditions	Conv. (%), TON/TOF	Stability evidence	Take home message	Ref.
8	Pt@UiO-66	Cinnamyl alcohol, O_2, toluene, 80°C, 9 h	100	Reuse, XRD, TEM, hot filtration	High activity under base-free conditions	[85]
9	Pd^0-in-UiO-67 (3.0 ± 0.5 nm)	Cyclic and linear alcohols, air, toluene, 80-120°C, 12-40 h	96-99	Reuse, XRD, TEM, hot filtration, AAS	High efficiency under air atmosphere and base-free conditions	[86]
10	Pd/DUT-67 (12-17 nm)	Benzylic alcohols, O_2, H_2O, base, 75°C, 8-24 h	7-100	Hot filtration, ICP, reuse, TEM, XRD, SEM	High catalytic activity and selectivity in neat water under an O_2 atmosphere	[87]
11	Pd-NPs/$Cu_2(BDC)_2$(DABCO) (2.9 ± 0.2 nm)	Benzylic alcohols, air, Na_2CO_3, 120°C, 24 h	93-99	ICP, reuse	Higher catalytic activity than common palladium salts	[88]
12	Pd/MIL-101 (2.5 ± 0.5 nm)	Cyclic and linear alcohols, O_2, toluene, 80°C, 0.5-24 h	96-99	Reuse, AAS, XRD,	High TOF of approximately 16900 h^{-1} under base-free conditions	[89]
13	Pd/Au@ZIF-8	Secondary alcohols, K_2CO_3, H_2O, O_2, 60°C, 24 h	0-5.9	-	Negligible catalytic activity: hydrophobic nature of the ZIF-8 shell around the NPs	[90]
14	Pd/MIL-101 (2-10 nm)	Benzyl alcohol, air, *o*-xylene, 85°C, 14 h	84.8	Reuse, hot filtration, ICP, TEM	Higher catalytic activities than the commercially available Pd-supported activated carbon	[91]

ZIF-8: $Zn(MeIM)_2$, MeIM: 2-methylimidazole, MOF-177: $[Zn_4O(BTB)_2]_8$: BTB:1,3,5-benzenetribenzoic acid, DUT-67: DUT, Dresden University of Technology, $[Zr_6O_6(OH)_2(tdc)_4(CH_3COO)_2]$, tdc: 2,5-thiophenedicarboxylate, DABCO: 1,4-diazabicyclo[2.2.2]octane.

Figure 14. Schematic illustration of the synthesis of Au@UiO-66 (NH_2) material. Reprinted from Ref. [78] with permission.

by Chen et al. in which a Pd^0-in-UiO-67 was synthesized by encapsulation of the Pd precursors through ligand design prior to the MOF assembling [86] (Figure 15). In contrast to the traditional impregnation method, the proposed synthetic route can avoid the diffusion resistance between external and internal surfaces, allowing metal precursors to be easily deposited into the MOF pores. The resulting Pd^0-in-UiO-67 was evaluated as a catalyst in the aerobic oxidation of a wide range of alcohols under air atmosphere and base-free conditions. Very high conversion of alcohols to the corresponding aldehydes (> 96%) was observed within 20-40 h (Table 3, entry 9).

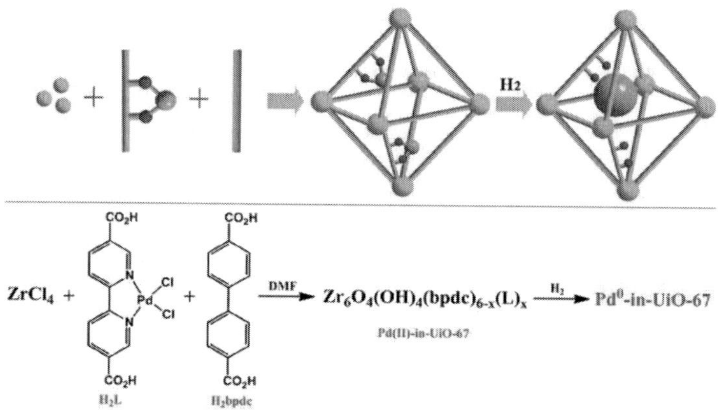

Figure 15. Schematic illustration of the synthesis of Pd^0-in-UiO-67 material. Reprinted from Ref. [86] with permission.

In contrast, the Pd/C having an almost identical particle size distribution displayed a conversion of 51% for cinnamyl alcohol with a selectivity of 72%. Furthermore, the Pd^0/UiO-67 prepared through the conventional impregnation method, in which the Pd NPs were mainly deposited on the external surface, presented 5% conversion in the absence of a base. The observed high activity of Pd^0-in-UiO-67 was attributed to the synergistic effects of nano-confinement and the electron donating ability of the UiO-67 framework. More specifically, the organic ligands donate electrons to the Pd surfaces through coordination or π-bond interactions possibly facilitating the formation of anionic Pd and thereby favouring O_2 activation on the Pd surfaces.

1.4. Direct Synthesis of Modified MOFs

In section 1.3, we presented the applicability of MOFs to incorporate a large variety of active sites into their frameworks by utilizing the chemical modification of the framework after synthesis (post-synthetic modification). Another synthetic approach to obtain advanced MOF materials suitable for more specialized applications is the introduction of active sites directly into the framework such as in situ embedding of NPs and pre-functionalization of linkers. The direct synthesis of MOFs using pre-functionalized ligands is challenging as the decomposition of the transition metal complexes under the applied reaction conditions can occur. Furthermore, it is essential to find the appropriate reaction conditions to form a particular MOF structure. As a result, very few papers report on the modification of MOFs using the pre-functionalized method, whereas there are many studies that consider the classical post-synthetic modification route to incorporate functional groups. As described in Section 1.3, Abednatanzi et al. [62] and Chen et al. [86] demonstrated the advantages of the direct synthesis approach using pre-functionalized linkers compared to the traditional post-synthetic modification. Other examples of the direct synthesis of modified MOFs are presented in Table 4. Besides the use of pre-functionalized linkers, active sites can be incorporated into the networks by mixing them with MOF precursors. In this regard, the self-assembly of DUT-5 around the in situ formed Pt NPs surface is obtained through the direct mixing of H_2PtCl_6 and MOF precursors without any additional protective agents and reductants

[92] (Figure 16). The resulting Pt@MOF nanocomposites display great stability, significantly enhanced catalytic activity and selectivity as compared to the commercial Pt/C in liquid phase aerobic oxidation of alcohols under mild base-free conditions (Table 4, entry 1).

Table 4. Catalytic activities of modified MOFs through direct synthesis approaches

Entry	Catalytic MOF	Substrate + Conditions	Conv. (%), TON/TOF	Stability evidence	Take home message	Ref.
1	Pt@DUT-5	Benzylic, cyclic and aliphatic alcohols, O_2, toluene, 80 or 100°C, 15-40 h	93-99	Reuse, XRD, N_2 sorption, TEM, AAS	Enhanced catalytic activity and selectivity compared to the commercial Pt/C	[92]
2	Fe_3O_4/Fe-MIL-101	Benzylic and cyclic alcohols, O_2, TEMPO, CH_3CN, KNO_2, 75°C, 14 h	0-99	Hot filtration, reuse, ICP	Broad substrate scope	[93]
3	Pd/Cu-MOF	Benzyl alcohol, O_2, 100°C, 10 h	~ 99, 10000	-	Extremely active: nanocatalysts of small size and the hierarchical porosity of the support	[94]

DUT-5: (Al(OH)(bpdc), bpdc: 4,4'-biphenyldicarboxylate.

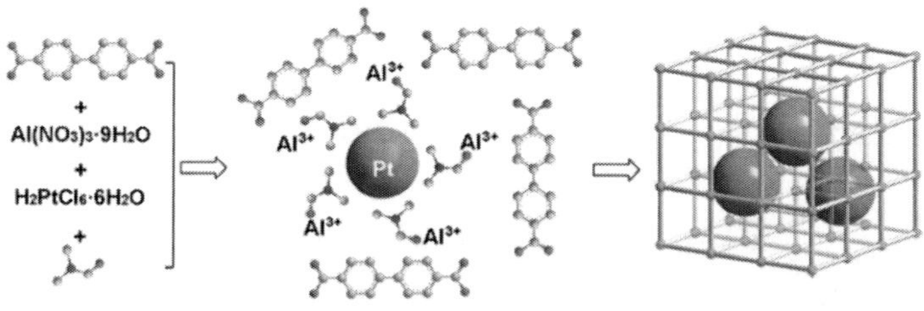

Figure 16. Schematic illustration of the synthesis of in situ one-step synthesis of Pt@DUT-5 (gray: C, red: O, blue: N, orange: Pt clusters or NPs, cyan: Al-based SBU). Reprinted from Ref. [92] with permission.

1.5. Synthesis of MOF Composites

The composite materials are the compounds made from two or more constituents, which exhibit new properties that are superior to those of the individual components. Combining MOFs and functional materials is an elegant approach to the creation of new multifunctional composites in order to merge the merits and decrease the shortcomings of both the components. In MOF composites the structural diversity and flexibility, high porosity of MOFs and optical, magnetic, catalytic properties of functional materials efficiently link together to introduce new physical and chemical properties. A large number of active components such as metal oxides, metal nanoparticles, quantum dots, graphene/graphene oxide, polymers and carbon nanotubes have been used in the design of MOF-based composites. The remarkable features and synergistic effect of the MOF composites have been successfully applied for a wide range of applications, particularly for catalysis including alcohol oxidation reactions.

1.5.1. MOF-Metal Oxide Composites

In Table 5 a series of MOF composites with different active species for the alcohol oxidation are summarized. As can be seen from this Table, metal oxide/MOF composites are one of the most studied catalysts in the oxidation reaction.

As an interesting example of metal oxide/MOF composites, amorphous TiO_2 within the prepared mesoporous HKUST-1 applied for selective aerobic oxidation of benzylic alcohols under sunlight irradiation [97]. In this regard, the ordered mesoporous HKUST-1 was synthesized via a surfactant-assisted templating method and then TiO_2 coated into the MOF via the layer-by-layer method. The formed composite showed moderate to high yields for benzylic alcohols oxidation in CH_3CN at room temperature using O_2 as the oxidant (Table 5, entry 3). The poor transformation of the alcohol in the optimal reaction condition for the bare MOF and amorphous TiO_2 revealed

Table 5. Catalytic activities of MOF composites for alcohol oxidation

Entry	Catalytic MOF	Substrate + Conditions	Conv. (%), TON/TOF	Stability evidence	Take home message	Ref.
1	Au@MOF-5 Au/ZnO@MOF-5 Au/TiO_2@MOF-5	Benzylic alcohols, oxygen, K_2CO_3, methanol, 30 min, 80°C	48-72	-	Homogeneous distribution of Au NPs with a size of ca. 1–3 nm	[95]
2	MOF/$BiFeO_3$	Benzylic alcohols, CH_3CN, H_2O_2, 45 min, 80°C	50-98	Reuse, XRD	Synergistic effect of $BiFeO_3$ and MOF	[96]
3	HKUST/TiO_2	4-methylbenzyl alcohol, CH_3CN, sunlight irradiation, O_2, 25-35°C, 8 h	6-100	Reuse, IR, XRD, TGA	Superior activity of HKUST/TiO_2 compared to the both HKUST and TiO_2	[97]
4	Fe_3O_4/MIL-101	Benzylic alcohols, CH_3CN, H_2O_2, reflux, 1-8 h	40-98	Reuse	Magnetically reusable catalyst	[98]
5	Fe_3O_4/MIL-101	Benzylic alcohols, TBHP, 70°C, 12 h	44-79.8	Reuse, XRD	Higher catalytic activity compared to Fe_3O_4, solvent-free catalysis	[99]
6	Fe_3O_4/$Cu_3(btc)_2$	Alcohols, CH_3CN, O_2, Na_2CO_3, TEMPO, 75°C, 6-12 h	2-99	Hot filtration, ICP, reuse, TEM, XRD	Magnetic recovery	[100]
7	Fe_3O_4@P4VP@ ZIF-8	Benzylic alcohols, O_2, CH_3CN, base, TEMPO, 70°C, 12 h	5-99	Reuse, XRD, TGA, IR, TEM	Excellent recovery properties	[101]
8	Fe_3O_4@P4VP@MIL -100(Fe)	Alcohols, CH_3CN, O_2, TEMPO, KNO_2, 75°C, 12 h	22-99	Reuse, IR, XRD, hot filtration	Magnetic reusability	[102]
9	GO/UiO-66-NH_2	Benzyl alcohol, benzotrifluoride, O_2, room Temp, 8 h, irradiation (λ> 420 nm)	2-43	Reuse	Higher efficiency than UiO-66-NH_2	[103]
10	Eu-Based MIL-LIC/GO	Benzyl alcohol, H_2O/CH_3CN, LED light, 3-6 h, 20°C	5-85	-	Water is used as the oxygen source	[104]

Entry	Catalytic MOF	Substrate + Conditions	Conv. (%), TON/TOF	Stability evidence	Take home message	Ref.
11	ZIF-67@GO	Benzyl alcohol, O_2, DMF, 100°C, 8 h	2-89.5	Reuse	Magnetic separation	[105]
12	$(py_3tren)Co_2$@NU-1000@SiO_2 $(py_3tren)AlCo$@NU-1000@SiO_2	Benzyl alcohol, TBHP, CD_3CN, 80°C, 2 h	52.8 42.4	Reuse	Higher efficiency than both NU-1000@SiO_2 and NU-1000	[106]
13	Pd@MIL-88B-NH_2@nano-SiO_2	Secondary benzylic alcohols, *p*-xylene, air, 10-68 h, 150°C	~ 0-98	Reuse, ICP, XRD, TEM	Aerobic oxidation under batch and continuous flow	[107]
14	ZIF-9@Zeolite	Benzylic alcohols, water, H_2O_2, 28-33 h, 80°C	80-98	Reuse, IR	Broad substrate scope	[108]
15	CdS–UiO–66 (NH_2)	Benzylic alcohols, benzotrifluoride, O_2, under visible-light irradiation (λ> 420 nm), 4 h	~ 21-63	Reuse, XRD	Higher photocatalytic activity than CdS and UiO–66(NH_2): synergistic effect of the effective transportation of the photogenerated electron–hole pairs	[109]
16	PEG/ $Cu_3(BTC)_2$	Benzyl alcohol, O_2, TEMPO, Na_2CO_3, CH_3CN, 6 h, 75°C	99	Reuse, TEM, XRD, hot filtration	Higher reusability and stability compared to the commercial $Cu_3(BTC)_2$	[110]
17	CCSs@HKUST-1	Benzyl alcohol, TEMPO, Na_2CO_3, CH_3CN, 70°C, 22 h	71.8	-	Better catalytic activity than the pure HKUST-1	[111]
18	Cu@C (pyrolysis of $Cu_3(BTC)_2$)	Benzylic alcohols, TEMPO, N-methylimidazole, O_2, CH_3CN, 70°C, 9 h	10-99	Leaching, reuse	Broad scope of substrates	[112]
19	Co-CoO@N-doped carbon (pyrolysis of ZIF-67)	Benzyl alcohols, O_2, CH_3OH, K_2CO_3, 80°C, 24 h	85-100	Reuse, leaching, hot filtration, TEM	Facile separation, broad scope of substrates, synergistic effect between Co and CoO NPs	[113]

Table 5. (Continued)

Entry	Catalytic MOF	Substrate + Conditions	Conv. (%), TON/TOF	Stability evidence	Take home message	Ref.
20	Co@C-N (thermolysis of ZIF-67)	Aromatic Alcohols, air, CH_3OH, *n*-hexane, 25°C, 96 h	80-99	Reuse, leaching, hot filtration, TEM, XRD	Oxidative Esterification, base- free, broad scope of substrates, reactivation in H_2 at 400°C	[114]
21	Co@C-N (by pyrolysis of core-shell ZIF-67@ZIF-8 precursors)	Alcohols, H_2O, air, 110°C, 1 h	69-99	Reuse, XRD, TEM	Atmospheric air as the sole oxidant, magnetic reusability, broad scope of substrates	[115]

P4VP: poly(4-vinylpyridine), Eu-MIL-LIC: Europium 2-Aminoterephthalic acid MOF, py$_3$tren: *N,N,N*-tris(2-(2-pyridylamino)ethyl)amine, MIL-88B: Chromium 2-Aminoterephthalic acid MOF, PEG: Poly(ethyleneglycol), CCSc: carboxyl rich carbon spheres.

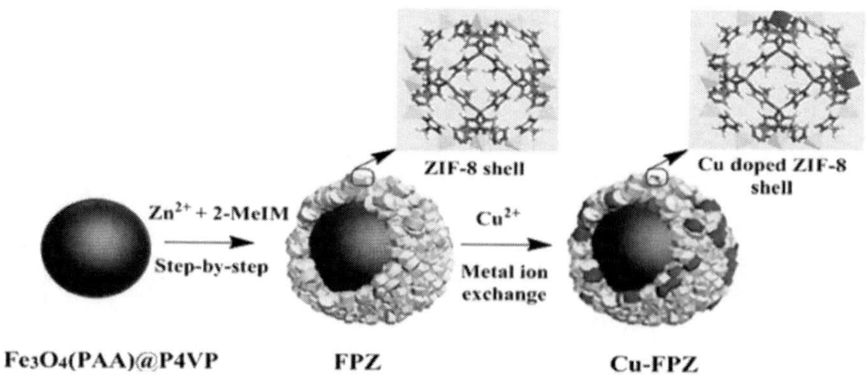

Figure 17. Schematic illustration of the magnetic core–shell copper-doped Fe_3O_4@P4VP@ZIF-8 catalyst preparation. Reprinted from Ref. [101] with permission.

the superior properties for the TiO_2/MOF composite. The synergistic effect of both a photoresponse ordered meso-MOF and amorphous titania is responsible for the observed uncommon reactivity. Another composite system was developed based on hierarchical magnetic microspheres of Cu-doped Fe_3O_4@P4VP@ZIF-8 [101]. The composite was prepared firstly by copolymerization of the 4-vinylpyridine, divinyl-benzene and polyvinyl pyrrolidone (PVP) on the surface of Fe_3O_4. Then the ZIF-8 MOF shell was grown on the surface of the monodispersed Fe_3O_4@P4VP core-shell. Finally, the Zn^{2+} ions in the ZIF-8 framework were replaced partially with the catalytically active Cu^{2+} ions (Figure 17). The high magnetic response of Fe_3O_4 core provides catalyst separation possibility by applying a magnetic field. Polymeric layer assisted the MOF formation and stabilization on the core-shell structure. The copper-doped Fe_3O_4@P4VP@ZIF-8 composite demonstrated 99% yield of benzaldehyde in the oxidation of benzyl alcohol using TEMPO as the co-catalyst and $NaHCO_3$ as the base in acetonitrile at 60°C with oxygen as the oxidant (Table 5, entry 7). The turnover number and turnover frequency of the magnetic core-shell were significantly improved in comparison to many other systems. The recycled composite catalyst could be successfully reused for 15 times and maintained 96% of its yield of benzaldehyde.

1.5.2. MOF-Graphene Composites

Graphene oxide (GO) possess many unique properties, including high mechanical strength, chemical stability and low electrical resistivity, which are desirable to integrate with other functional materials to construct composites. Various functional groups such as hydroxyl, epoxide and carboxylate groups on the plane and at the edges of the GO layers and aromatic sp^2 domains can participate in the bonding interactions in MOFs. The incorporation of GO into the frameworks of MOFs can improve the stability, porosity, electrical conductivity and photocatalytic properties of MOFs. In this regard, a layered nanoarchitecture composed of photoactive UiO-66-NH_2 MOF and graphene was fabricated (Figure 18) and its activity was tested in the oxidation of benzyl alcohol [103]. The GO/UiO-66-NH_2 composite was prepared via a noncovalent methodology for graphene functionalization with $pyrene^+$ combined with in situ self-assembling and a solvothermal procedure. The selective photocatalytic oxidation of benzyl alcohol can boost by GO/UiO-66-NH_2. Under the optimized reaction conditions, GO/UiO-66-NH_2 and pure UiO-66-NH_2 MOF as catalysts resulted in 18.6% and 11.6% of benzyl alcohol conversion respectively, under the oxygen atmosphere at room temperature and irradiation wavelength of $\lambda > 420$ nm (Table 5, entry 9). The authors suggested that the intimate contact between GO and MOFs through pyrene$^+$ is able to promote the separation and transfer of photogenerated charges which increases the catalytic activity.

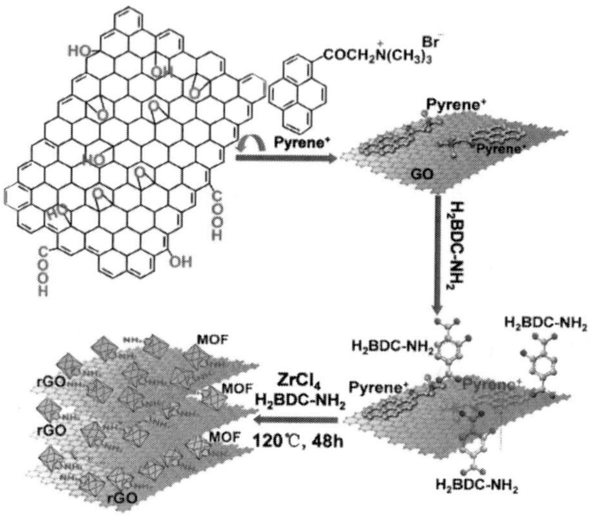

Figure 18. Schematic illustration of the stepwise fabrication procedures for the layered GO/UiO-66-NH_2. Reprinted from Ref. [103] with permission.

Recently, a newly designed Eu-based MIL-LIC/GO composite was utilized as a photocatalyst for the oxidation of benzyl alcohol using water as an oxygen source [104]. GO is considered as one of the most promising candidates for an ideal functional photocatalyst carrier and promoter due to its large specific surface area, high electrical conductivity and superior visible light adsorption. In the MIL@GO composite, GO can act as structure- directing agent in controlling the morphology. Moreover, reducing the recombination of photogenerated electron–hole pairs can promote the photocatalytic activity. The activity of this catalyst was examined in the oxidation of benzyl alcohol using CH_3CN and water as solvents and under LED light in N_2 atmosphere. MIL@GO showed higher photocatalytic activity (84.5% conversion) than that of the bare MIL-LIC-1(Eu) (Table 5, entry 10). The synergistic effect between GO and MIL-LIC-1(Eu) can be attributed to GO ability to inhibit the recombination of photogenerated electron–hole pairs. The control experiment with the EDTA (as a hole scavenger) and *tert*-butyl alcohol (as a $OH^{\bullet}$ scavenger) revealed that holes and $OH^{\bullet}$ are the active species in the photocatalytic oxidation.

1.5.3. MOF-Silica Composites

Silica nanoparticles and nanostructures with high porosity, stability, dielectric properties and possibilities for introducing multiple functionalities are broadly employed in applications like catalysis, separation and drug release. Integrating silica nanomaterials and MOFs into core-shell nanostructures is an important and effective way to further improve the properties and introduce new functionalities. There are two different approaches to design MOF-silica composites including SiO_2@MOFs and MOFs@SiO_2. In the SiO_2@MOF composites, the silica nanomaterials can incorporate inside the pores of MOFs or a MOF can grow on a silica sphere. While in the MOFs@SiO_2 case, a silica shell can employ as a surface coating on MOFs. One of the most important advantages of these composites is their enhanced affinity to hydrophilic molecules. This results in an improved water dispersibility as well as stability. Moreover, a better control over the size and shape of MOFs can be obtained.

The fabrication of a palladium nanocatalyst protected by an innovative double-layered MOF@silica support and its application in the aerobic oxidation of benzylic alcohols has been reported by Pascanu and co-worker [107]. The coating of Pd@MOF with a shielding layer of mesoporous silica drastically increases the durability and recyclability of the catalyst under batch and continuous flow condition (Figure 19). Pd@MIL-88B-NH_2@nano-SiO_2 was used as a solid catalyst for the aerobic oxidation of 1-phenylethanol using air as the oxidant in *p*-xylene at 150°C, which resulted in 98% yield of the related aldehyde under optimal conditions (Table 5, entry 13). A wide variety of substituted secondary benzylic alcohols were oxidized in high yields under identical conditions. Under batch conditions, by using a "teabag" approach as an efficient recycling protocol, the catalyst maintained its activity with almost no decrease in four consecutive cycles. More interestingly, the catalyst was used for 7 days of continuous operation without significant loss of its activity under continuous flow.

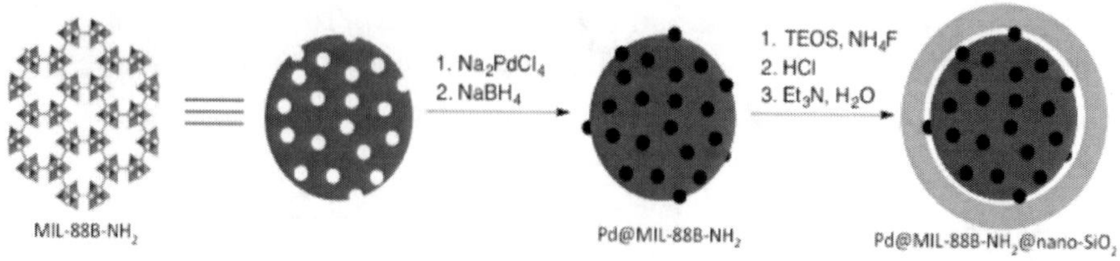

Figure 19. Graphical illustration of the stepwise synthesis of Pd@MIL-88B-NH_2@nano-SiO_2. Reprinted from Ref. [107] with permission.

1.6. MOFs for Photocatalytic Oxidation

One of the main targets in chemistry is the development of renewable and green energy based processes. Photocatalytic organic transformation by employing light energy is a good candidate for many different environmentally benign organic processes. The integration of the photoactive sites into the modular nature of MOFs could result in their synergistic and superior functions in key areas such as optoelectronics, light harvesting, and photoinduced heterogeneous catalysis. The metal nodes, organic moieties and the cavity in MOFs can all be used as photosensitizer agents in photocatalysis (Table 6).

Recently, the Pt nanocrystals (NCs) and photosensitizing porphyrinic MOF (PCN-224) were assembled to obtain Pt/PCN-224(M) composites (M represents the metal ion residing in the center of porphyrin ligand: Zn, Ni, Co, Mn and 2H) [118]. The catalytic performance of these Pt/PCN-224(M) composite materials were studied in the photo-oxidation of benzyl alcohol using O_2 at ambient temperature in H_2O. The Pt/PCN-224(Zn) photocatalyst was able to achieve up to 99% conversion of benzyl alcohol with high selectivity (100%) (Table 6, entry 3). Furthermore, Pt/PCN-224(Zn) was also an active photocatalyst to transform diverse substituted benzyl alcohols and heteroaromatic alcohols into their corresponding carbonyl compounds with quantitative conversions (>99%) as well as complete selectivity towards their corresponding aldehyde (100%). From the catalysis viewpoint, the most interesting feature of the Pt/PCN-224(Zn) photocatalyst system is

Table 6. Catalytic activities of MOFs for alcohol photo-oxidation

Entry	Catalytic MOF	Substrate + Conditions	Conv. (%), TON/TOF	Stability evidence	Take home message	Ref.
1	Mixed-linker UiO-66–NH_2-X (X: H, F, Cl, Br)	Benzyl alcohol, mesitylene, O_2, toluene, 26 W Helical bulb irradiation, 80°C, 24 h	11.7-53.9	-	Higher efficiency compared to the UiO-66-NH_2	[116]
2	Ti-based MIL-125 (aminated linkers: 0, 20,46, 70 and 100%)	Benzyl alcohol, mesitylene, O_2, toluene, 300 W Xe Arc lamp (λ=325, 415, 450), room Temp., 24 h	276 mmol h^{-} mg^{-}	Reuse	Maximized photocatalytic activity with aminated-functionalized linkers, solvent-free catalysis	[117]
3	Pt/PCN-224 (M) (M: Zn, Ni, Co, Mn, 2H)	Aromatic alcohols, water, O_2, ambient temperature, 300 W Xe lamp (λ=400 nm) irradiation, 40-300 min	99	Reuse, XRD, TEM, XPS	Synergetic photothermal effect and singlet oxygen production	[118]
4	PCN-22	Benzylic alcohols, CH_3CN, TEMPO, O_2, visible light, 2 h	25	Reuse, XRD	Synergetic effect of combining inorganic and organic components	[119]
5	MIL-125(Ti)-AM-Eu	α-phenethyl alcohol, DMF, O_2, 600 W ultraviolet lamp, room temp., 6 h	20.4	Reuse, ICP, XRD,	Higher activity compared to MIL-125(Ti)-NH_2	[120]
6	Au@ZIF-8	Benzyl alcohol, CH_3CN, O_2, under Xe lamp irradiation (λ=400 nm), 24 h	25.8-51.6	-	Reduced activity due to the size selectivity of ZIF-8	[121]
7	UiO-66(NH_2)	Alcohols, benzotrifluoride, O_2, visible-light irradiation (λ ≥ 420 nm), 60°C, 4 h	8.5-35.5	Reuse	Tuning photo-absorption edge to the visible light: introducing the amino group (-NH_2)	[122]

Table 6. (Continued)

Entry	Catalytic MOF	Substrate + Conditions	Conv. (%), TON/TOF	Stability evidence	Take home message	Ref.
8	CdS–MIL-100(Fe)	Benzyl alcohol, toluene, Xe lamp irradiation (λ > 420 nm), 5 h	54.3	Reuse	Higher activity compared to CdS and MIL-100(Fe)	[123]
9	QUI-MIL-125(Ti)	Benzyl alcohol, CH_3CN, white-light LED lamp (λ = 420–780 nm), O_2, room Temp., 40 h	46-99	Reuse, IR	High catalytic activity: Reducing the band gap through conjugated linker engineering	[124]
10	PANI/FeUiO-66	Benzylic alcohols, ethyl acetate, O_2, Xe lamp irradiation (λ > 400 nm), 5 h	10-100	Reuse, IR, XRD	Superior photocatalytic efficiency compared to UiO-66, FeUiO-66 and PANI/UiO-66	[125]
11	Ni-doped NH_2-MIL-125(Ti)	Aromatic alcohol, benzotrifluoride, O_2, under Xe lamp irradiation (λ > 420 nm), room Temp., 10 h	10.5-47.4	Reuse, XRD, BET	Significant enhanced photocatalytic activity by doping Ni NPs into NH_2-MIL-125(Ti)	[126]

MIL-125: $Ti_8O_8(OH)_4(bdc)_6$, PCN-224: $Zr_6(\mu_3\text{-}O)_4(\mu_3\text{-}OH)_4(OH)_6(H_2O)_6(TCPP)_{1.5}$ (TCPP = tetrakis(4-carboxyphenyl)-porphyrin), PCN-22: $[Ti_7(\mu_3\text{-}O)_2O_2(OH)_2](TCPP)_3$ $(DEF)_2$ (DEF: N,N'-diethylformamide), QUI: 2-quinolinecarboxaldehyde, PANI: polyaniline.

the utilization of singlet oxygen for selective oxidation of aromatic alcohols to aldehydes. The high visible-light harvesting power of both Pt NCs and PCN-224(Zn) synergistically activate molecular oxygen from its triplet ground state into the singlet excited states as a moderate and selective oxidant. In addition, the photothermal effect, electronic state of Pt surface, and diamagnetism of Zn^{2+} ion in the Pt/PCN-224(Zn) composite promote significantly the catalytic performance (Figure 20).

The fundamental steps in a photocatalytic reaction including (1) light absorption; (2) charge separation and transfer; and (3) redox reaction can be improved in MOF-based photocatalysts by employing different strategies such as ligand functionalization, direct integrating a sensitizer or quantum dots into the framework of MOFs as well as the use of ligands with large light harvesting. For example, a series of aromatic heterocycle-grafted MOFs were prepared by applying a post-grafting procedure through a Schiff base chemical reaction between the aldehyde and $-NH_2$ groups within NH_2-MIL-125(Ti) [124]. The photocatalytic activity of the obtained catalysts was studied in the oxidation of alcohols under visible-light irradiation. The conversion of benzyl alcohol was 69% with NH_2-MIL-125(Ti) as a catalyst. Under the same reaction condition all the aromatic heterocycle-grafted MOFs exhibit significantly higher photocatalytic activity than primary NH_2-MIL-125(Ti). Among the various catalysts examined, the NH_2-MIL-125(Ti) grafted by 2-quinolinecarboxaldehyde (QUI) and 3-thiophenecarboxaldehyde (THI) exhibited the highest catalytic activities resulting in 88 and 86% conversions of benzyl alcohol under identical conditions. The superior catalytic activity of the QUI-MIL-125(Ti) and THI-MIL-125(Ti) is due to their lower band gaps. The modification of NH_2-MIL-125(Ti) with QUI and THI efficiently expanded the π- delocalized system and promoted the light adsorption and charge carriers migration.

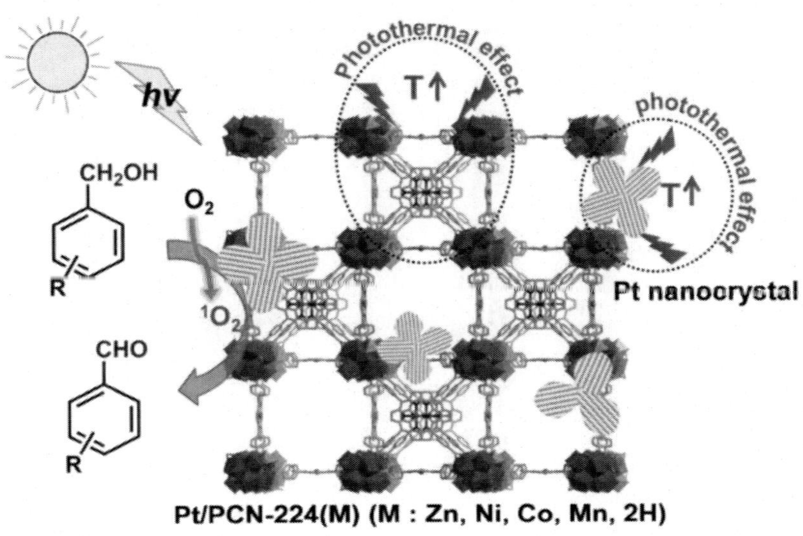

Figure 20. Graphical illustration showing the selective oxidation of alcohols over Pt/PCN-224(M) using O_2. Reprinted from Ref. [118] with permission.

2. General Introduction on Covalent Organic Frameworks (COFs) and Covalent Triazine Frameworks (CTFs)

A very novel class of materials, the Covalent Organic Frameworks or COFs, could be a perfect candidate to create heterogeneous oxidation catalysts. As already mentioned in the introduction, the "crystalline porous polymers" are very light and some are extremely stable.

First described by Omar Yaghi in 2005 [127], the COF family has been extended by several new structures, of which the Covalent Triazine Frameworks (CTFs) have shown to be remarkably stable, as they can withstand conditions of extreme acidic or basic pH and treatments up to 400°C. They were first reported in 2008 by Thomas and Antonietti [128].

CTF materials are prepared by the reversible triazine ring formation by polymerizing dicyanoaryls. The individual sheets then stack by pi-pi interactions into a hexagonal 2D layered structure. The synthesis approach towards CTF type COFs is shown in Figure 21.

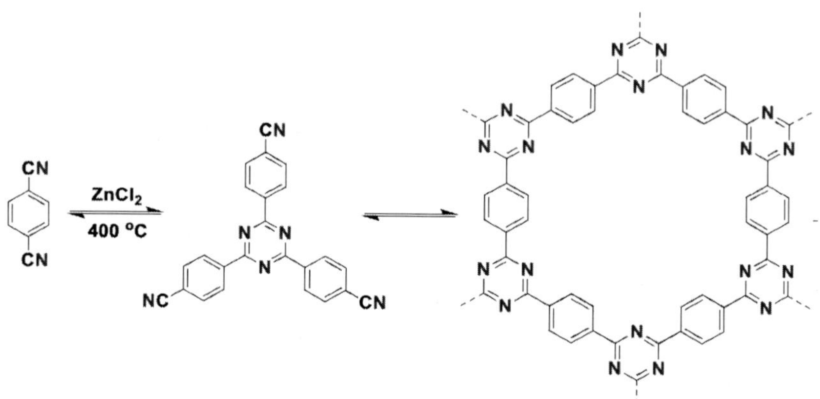

Figure 21. Schematic illustration for the synthesis of CTF-1.

The most significant challenge is to design robust synthesis procedures that do not need such a harsh condition. Enormous effort is recently made to construct CTFs under mild conditions, which benefit for further development and practical large-scale synthesis of CTFs [129-130].

Another type of COFs that are suitable for oxidation reactions is the so-called imine-COFs. The first imine-based COF was developed by Yaghi et al. and was denoted COF-300 [131]. This material was obtained via the dehydration of aldehydes and amines giving rise to Schiff-base type linkages. Another hexagonal imine-COF was prepared by Wang et al., denoted as LZU-1 [132]. The imine-based COFs are found to have comparable crystallinity as the boron-containing COFs. Moreover, the imine-based COFs are stable in most organic solvents and are insensitive to water [133]. Banerjee et al. [134-135] showed the complete stability of the imine-based COFs after being immersed in boiling water and in acidic and basic media (9 molar) for 7 days. Another ultrastable variant has been developed very recently, withstanding treatments in 12 M HCl and 14 M NaOH solutions for a week [136]. A typical synthesis approach for triazine-based COFs is shown in Figure 22.

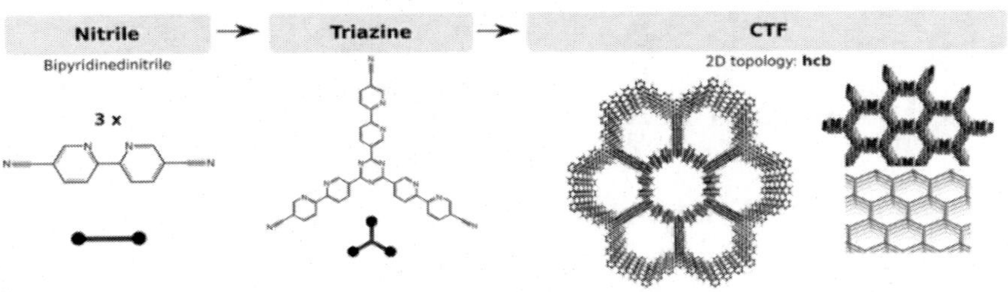

Figure 22. Typical synthesis approach for triazine-based COFs.

Next to thermal oxidation catalysts, COFs (and particular CTFs) are one of the most significant discoveries pertaining to heterogeneous photo catalysis because [137]:

(1) They are composed of molecular building blocks and can be tuned to optimize light harvesting, charge separation, charge transport and electrocatalysis.

(2) They have permanent nanometer-sized pores that can also be tuned. Highly porous materials with very large surface areas (> 1500 m^2/g) provide a huge pool of adsorption sites for the VOCs, and a rapid diffusion of charges.
(3) The photoactive building blocks can be locked in a rigid structure, enhancing the lifetime of the excited states by preventing deactivation through collision.
(4) The crystalline or semi-crystalline nature of these materials facilitates charge transport and minimizes charge trapping at defect sites.
(5) They are very stable and robust.
(6) They have a very low density, much lower than MOFs as they are solely composed of light elements, C, N, O, H.

The exceptional mix of the solid-state character of the CTFs, combined with tunability, porosity and crystallinity offers great and largely unexplored opportunities in photocatalytic research.

2.1. Designing COFs and CTFs for Selective Alcohol Oxidation

In previous sections, we demonstrated the unique features of MOFs for alcohol oxidation. However, it is important to address some of their limitations which restrict their use in heterogeneous catalysis. A majority of MOFs have a limited thermal and chemical stability and their stability under the reaction condition is still unsatisfactory. This limitation results in a poor catalytic recyclability which is very important in heterogeneous catalysis. Moreover, a series of steps is required for the synthesis of MOFs to provide an active network free of solvent molecules. Therefore, different activation procedures are needed to remove the solvent molecules that may be toxic and dangerous to the environment. On the other hand, in some cases, the structure collapses upon the activation. Nevertheless, in contrast to MOFs, COFs possess a higher thermal and chemical stability allowing pre- or post-synthetic modifications for specific applications without introducing a significant change in the porosity and loss in crystallinity. At this moment, the application of COFs and CTFs as selective oxidation catalysts is still in an initial stage, nevertheless, the obtained data so far exhibit promising results in terms of both sustainability and activity.

As an elegant example for a greener oxidation pathway, a thiophene-based CTF has been developed as a metal-free and photoactive catalyst for the selective aerobic oxidation of alcohols [138]. It is known that the ionothermal synthesis of CTFs leads to carbonization of the framework which results in unfavorable energetic bands for applying as a photocatalyst. Within this context, a mild synthetic route using trifluoromethanesulfonic acid (TfOH) as the catalyst to obtain a colorful material applicable for the photocatalytic reactions is applied by Huang and co-authors. As shown in Figure 23, the cyclization polymerization of the 2,5-dicyanothiophene (DCT) linker under TfOH vapor, a thiophene-containing CTF (CTF-Th) was directly synthesized onto mesoporous silica SBA-15, providing the photoactive CTF-Th@SBA-15 catalyst. The presence of the thiophene units into the CTF provides a low HOMO level that is desirable for a great photooxidation potential. Furthermore, applying mesoporous silica as the support for CTF could increase the active area in the photocatalytic reaction.

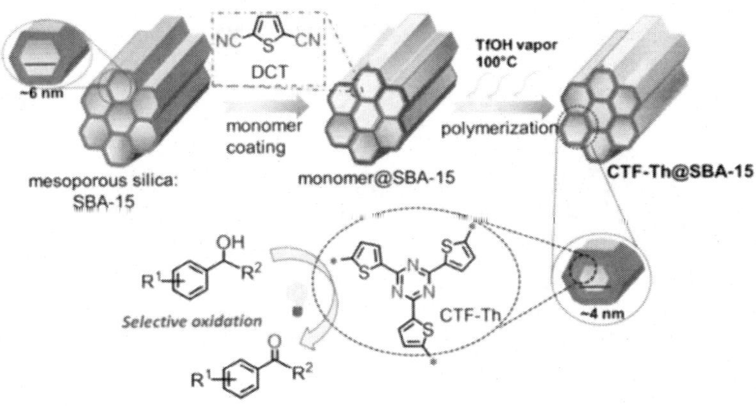

Figure 23. Schematic representation of thiophene-based CTF formation pathway on SBA-15. Reprinted from Ref. [138] with permission.

Table 7. Catalytic activities of COFs and CTFs in the selective oxidation of alcohols

Entry	Catalytic MOF	Substrate + Conditions	Conv.(%), TON/TOF	Stability evidence	Take home message	Ref.
1	CTF-Th@SBA-15	Benzylic alcohols, benzotrifluoride, O_2, blue light (λ = 460 nm), room Temp., 4 h	78-99	Reuse, IR, UV-vis	Highly efficient metal-free photocatalyst	[138]
2	COF-Am	Alcohols, CH_3CN, TBHP, 60°C, 14 h	0-95	Reuse, XRD, IR	Metal-free catalyst	[139]
3	Pd/CTF	Benzyl alcohol, cyclohexane, O_2, 80°C, 2 h	98	Reuse, ICP, hot filtration	Higher dispersion and stability of Pd NPs compared to AC and CNT	[140]
4	Pd@NHCS	Cinnamyl alcohol, toluene, air, 80°C, 10 h	99	-	Base-free, Higher activity compared to Pd/C and Pd/N-C	[141]
5	Pd/COP-4	Benzyl alcohol, O_2, 160°C, 1 h	32	ICP	Solvent-free selective oxidation	[142]
6	Pd/CMP-OH-H_2	Benzyl alcohol, toluene, O_2, 150°C, 4 h	74	-	Hydroxy-substituted CMP: higher catalytic activity than CMP-H_2	[143]
7	Ru-CTF	Benzyl alcohol, water, 0.1 M $HClO_4$, 1.4 V, 24 h	36.7	-	Electrooxidation of benzyl alcohol in water over the oxygen evolution reaction	[144]

CTF-Th: thiophene based covalent triazine framework, SBA: mesoporous silica, COF-Am: two-dimensional amide linked covalent organic framework, AC: activated carbon, CNT: carbon nanotubes, NHCS: nitrogen doped hollow carbon spheres, COP: covalent organic polymer, CMPs: conjugated microporous polymers.

The CTF-Th@SBA-15 catalyst showed a high conversion of benzyl alcohol (99%) towards benzaldehyde under blue light irradiation and oxygen at room temperature. After 4 h benzaldehyde was obtained in 99% conversion and 99% selectivity that is comparable with reported metal and non-metal catalytic systems (Table 7, entry 1). A series of control experiments were done to obtain insights into the reaction mechanism. More specifically, the addition of benzoquinone (superoxide scavenger), ammonium oxalate (hole scavenger) and NaN_3 (singlet oxygen scavenger) resulted in a decreased conversion of benzyl alcohol. These observations confirmed that the oxidation is associated with a combined action of superoxide ($^{\bullet}O_2^-$) and singlet oxygen ($^1O_2$) which are produced under the light irradiation. The proposed reaction mechanism (Figure 24) is based on molecular oxygen activation into its activated forms ($^{\bullet}O_2^-$ and $^1O_2$) by the photogenerated electron under visible light irradiation followed by the activation of benzyl alcohol and formation of benzaldehyde.

Another example of a metal-free catalyst based on COFs is developed via the condensation of 1,3,5-benzenetricarbonyl trichloride and *p*-phenylenediamine producing a two dimensional amide-COF [139]. This metal-free heterogeneous catalyst in the presence of TBHP has employed for the selective oxidation of benzyl alcohols to the corresponding benzaldehydes with yields up to 95% and 100% selectivity (Table 7, entry 2). The proposed mechanism by authors suggests that the reaction undergoes *via a* free radical mechanism assisted by the conjugated π-cloud and presented amide functionality in the COF (Figure 25).

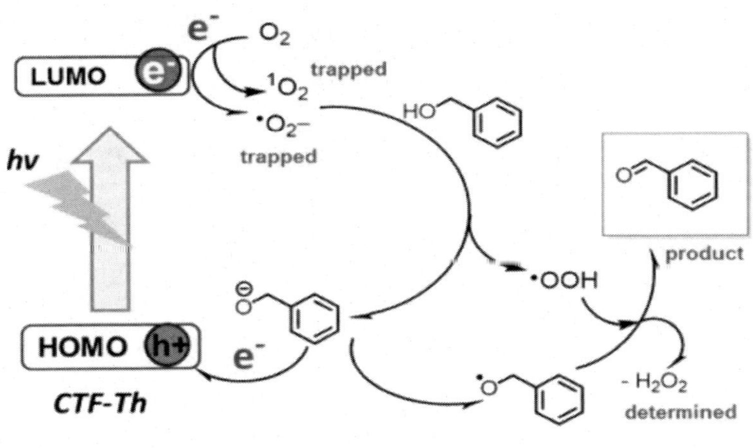

Figure 24. The proposed reaction mechanism for the selective oxidation of alcohols using CTF@SBA-15 as a metal-free photocatalyst. Reprinted from Ref. [138] with permission.

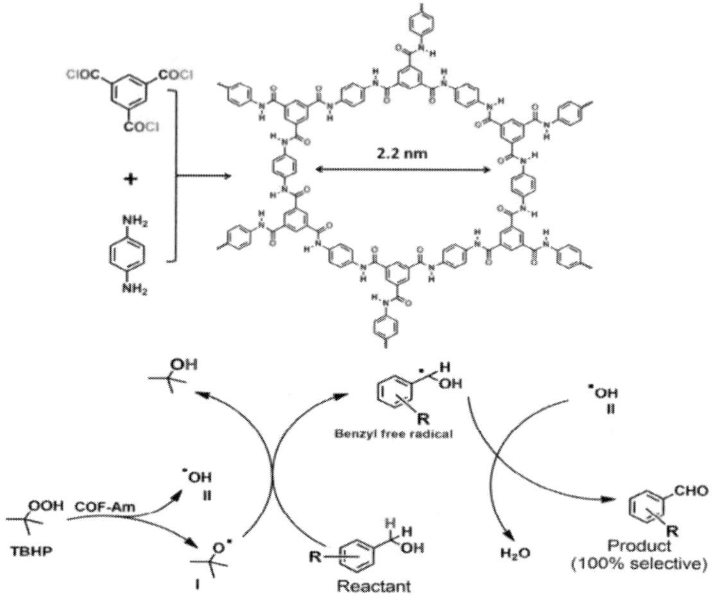

Figure 25. Schematic representation of the COF-Am synthesis and the proposed free radical mechanism for the oxidation reaction. Reprinted from Ref. [139] with permission.

Rather than the use of COFs and CTFs as metal-free catalysts, they can be used as support for a variety of metal complexes and NPs. Chan-Thaw et al. reported highly dispersed Pd NPs on a nitrogen-rich CTF, which showed good activity in the oxidation of benzyl alcohol to benzaldehyde. Compared to the nitrogen-doped carbon nanotubes (CNTs), CTFs are more efficient in stabilizing Pd NPs due to the presence of well-distributed nitrogen functionalities. In this regard, Pd/CTF showed higher activity than Pd/AC and Pd/CNTs due to the smaller size of NPs. However, the initial activity of Pd/nitrogen-doped CNTs was similar to Pd/CTF confirming the efficiency of nitrogen groups in the support for the immobilization of Pd NPs. The different number of N groups in N-CNTs and CTF (2% and 9% respectively) caused to the higher activity of CTF which is another proof to support the conclusion that the presence of nitrogen groups has a positive effect on the catalytic activity in alcohol oxidation (Table 7, entry 3) [140].

CONCLUSION

From this chapter, it can be concluded that there is an enormous growth of reports on MOFs and COFs in oxidation catalysis.

Regarding the MOFs, different types of alcohols have already been examined in which MOFs were examined as such (e.g., MIL-100, Cu-BTC, MIL-47.) or served as a matrix for the encapsulation or post-modification with a homogeneous catalyst. In many of these reports, the MOFs showed promising results in terms of activity, selectivity and stability. Nevertheless, to be able to compete with the industrial catalysts and to explore their industrial relevance, it is important that longer and more catalytic cycles are carried out in contrast to the few cycles (3-10 runs) presented now. At this moment, very few papers address the comparison of the MOF material with catalysts used in industry (e.g., zeolites, zeotypes, silica-based materials). Most of the reports compare the catalytic activity of the MOF with the homogeneous counterpart or metal oxides.

Within the concept of green chemistry, more efforts should be carried out to perform the oxidation reactions with clean oxidants like H_2O_2 or oxygen. Their use is probably hampered due to the very few MOFs which are known to be stable in water or air. We witness therefore a shift in research focus: in contrast to the original aim to design MOFs with unsaturated metal sites for catalytic applications, research groups are performing nowadays more efforts to enhance the intrinsic stability of the framework. This can be done by selecting a more rigid SBU or by performing a post-modification of a known stable framework. Only a few very stable frameworks have been reported, they include amongst others the UiO-66 (Zr) and several variants for the MIL-101 and MIL-100 (Cr, Al).

COFs (c.q. the CTFs and the imine-COFs) are emerging as ultra-stable support materials, and some nitrogen rich COFs or CTFs have (photo)catalytic activity by themselves as well. Their lightness and low toxicity are two important assets. Contrary to activated carbons, they are highly tunable in pore size and functional groups.

The elephant in the room is the fact that the harsh ionothermal synthesis conditions of the CTFs (molten $ZnCl_2$ at 400°C for an extended period of time) renders the largest part of the CTFs in fact amorphous and many functional groups are lost. CTFs made in those conditions are black and cannot be used in photocatalysis. The most important challenge is to create robust synthesis routes that do not need such drastic conditions. Huge progress is recently made to synthesis CTFs under mild conditions that benefit for further development and practical large-scale synthesis.

Regarding the imine COFs, good 2D imine COFs have been made in the last couple of years. The 3D imine COFs still form a challenge, although important breakthroughs have emerged while we were even writing this chapter (JACS paper in ILs for 3D imine COF).

As imine COFs are in principle better to control than CTFs, are stable and can be made as 3D cage structures, we expect huge progress in this field in the coming years.

References

[1] Long, J. R.; Yaghi, O. M., The pervasive chemistry of metal-organic frameworks. *Chem Soc Rev* 2009, *38* (5), 1213-1214.

[2] Rowsell, J. L. C.; Yaghi, O. M., Metal-organic frameworks: a new class of porous materials. *Micropor Mesopor Mat* 2004, *73* (1-2), 3-14.

[3] Leus, K.; Liu, Y. Y.; Van Der Voort, P., Metal-Organic Frameworks as Selective or Chiral Oxidation Catalysts. *Catal Rev* 2014, *56* (1), 1-56.

[4] Farha, O. K.; Eryazici, I.; Jeong, N. C.; Hauser, B. G.; Wilmer, C. E.; Sarjeant, A. A.; Snurr, R. Q.; Nguyen, S. T.; Yazaydin, A. O.; Hupp, J. T., Metal-Organic Framework Materials with Ultrahigh Surface Areas: Is the Sky the Limit? *J Am Chem Soc* 2012, *134* (36), 15016-15021.

[5] Doonan, C. J.; Morris, W.; Furukawa, H.; Yaghi, O. M., Isoreticular Metalation of Metal-Organic Frameworks. *J Am Chem Soc* 2009, *131* (27), 9492-+.

[6] Eddaoudi, M.; Kim, J.; Rosi, N.; Vodak, D.; Wachter, J.; O'Keeffe, M.; Yaghi, O. M., Systematic design of pore size and functionality in isoreticular MOFs and their application in methane storage. *Science* 2002, *295* (5554), 469-472.

[7] Czaja, A. U.; Trukhan, N.; Muller, U., Industrial applications of metal-organic frameworks. *Chem Soc Rev* 2009, *38* (5), 1284-1293.

[8] Tonigold, M.; Lu, Y.; Bredenkotter, B.; Rieger, B.; Bahnmuller, S.; Hitzbleck, J.; Langstein, G.; Volkmer, D., Heterogeneous Catalytic Oxidation by MFU-1: A Cobalt(II)-Containing Metal-Organic Framework. *Angew Chem Int Edit* 2009, *48* (41), 7546-7550.

[9] Ma, L. Q.; Abney, C.; Lin, W. B., Enantioselective catalysis with homochiral metal-organic frameworks. *Chem Soc Rev* 2009, *38* (5), 1248-1256.

[10] Silva, C. G.; Luz, I.; Xamena, F. X. L. I.; Corma, A.; Garcia, H., Water Stable Zr-Benzenedicarboxylate Metal-Organic Frameworks as Photocatalysts for Hydrogen Generation. *Chem-Eur J* 2010, *16* (36), 11133-11138.

[11] Gandara, F.; Puebla, E. G.; Iglesias, M.; Proserpio, D. M.; Snejko, N.; Monge, M. A., Controlling the Structure of Arenedisulfonates toward Catalytically Active Materials. *Chem Mater* 2009, *21* (4), 655-661.

[12] Ferey, G.; Mellot-Draznieks, C.; Serre, C.; Millange, F.; Dutour, J.; Surble, S.; Margiolaki, I., A chromium terephthalate-based solid with unusually large pore volumes and surface area. *Science* 2005, *309* (5743), 2040-2042.

[13] Chui, S. S. Y.; Lo, S. M. F.; Charmant, J. P. H.; Orpen, A. G.; Williams, I. D., A chemically functionalizable nanoporous material [Cu-3(TMA)(2)(H2O)(3)](n). *Science* 1999, *283* (5405), 1148-1150.

[14] Banerjee, M.; Das, S.; Yoon, M.; Choi, H. J.; Hyun, M. H.; Park, S. M.; Seo, G.; Kim, K., Postsynthetic Modification Switches an Achiral Framework to Catalytically Active Homochiral Metal-Organic Porous Materials. *J Am Chem Soc* 2009, *131* (22), 7524-+.

[15] Hong, D. Y.; Hwang, Y. K.; Serre, C.; Ferey, G.; Chang, J. S., Porous Chromium Terephthalate MIL-101 with Coordinatively Unsaturated

Sites: Surface Functionalization, Encapsulation, Sorption and Catalysis. *Adv Funct Mater* 2009, *19* (10), 1537-1552.

[16] Marx, S.; Kleist, W.; Baiker, A., Synthesis, structural properties, and catalytic behavior of Cu-BTC and mixed-linker Cu-BTC-PyDC in the oxidation of benzene derivatives. *J Catal* 2011, *281* (1), 76-87.

[17] Noei, H.; Amirjalayer, S.; Muller, M.; Zhang, X. N.; Schmid, R.; Muhler, M.; Fischer, R. A.; Wang, Y. M., Low-Temperature CO Oxidation over Cu-Based Metal-Organic Frameworks Monitored by using FTIR Spectroscopy. *Chemcatchem* 2012, *4* (6), 755-759.

[18] Kim, M.; Cahill, J. F.; Fei, H. H.; Prather, K. A.; Cohen, S. M., Postsynthetic Ligand and Cation Exchange in Robust Metal-Organic Frameworks. *J Am Chem Soc* 2012, *134* (43), 18082-18088.

[19] Gandara, F.; de Andres, A.; Gomez-Lor, B.; Gutierrez-Puebla, E.; Iglesias, M.; Monge, M. A.; Proserpio, D. M.; Snejko, N., A rare-earth MOF series: Fascinating structure, efficient light emitters, and promising catalysts. *Cryst Growth Des* 2008, *8* (2), 378-380.

[20] Farha, O. K.; Shultz, A. M.; Sarjeant, A. A.; Nguyen, S. T.; Hupp, J. T., Active-Site-Accessible, Porphyrinic Metal-Organic Framework Materials. *J Am Chem Soc* 2011, *133* (15), 5652-5655.

[21] Zhu, Q. L.; Li, J.; Xu, Q., Immobilizing Metal Nanoparticles to Metal-Organic Frameworks with Size and Location Control for Optimizing Catalytic Performance. *J Am Chem Soc* 2013, *135* (28), 10210-10213.

[22] Zhao, H. Y.; Gabbai, F. P., A bidentate Lewis acid with a telluronium ion as an anion-binding site. *Nat Chem* 2010, *2* (11), 984-990.

[23] Leus, K.; Liu, Y. Y.; Meledina, M.; Turner, S.; Van Tendeloo, G.; Van der Voort, P., A Mo-VI grafted Metal Organic Framework: Synthesis, characterization and catalytic investigations. *J Catal* 2014, *316*, 201-209.

[24] Bai, C. H.; Li, A. Q.; Yao, X. F.; Liu, H. L.; Li, Y. W., Efficient and selective aerobic oxidation of alcohols catalysed by MOF-derived Co catalysts. *Green Chem* 2016, *18* (4), 1061-1069.

[25] Enache, D. I.; Edwards, J. K.; Landon, P.; Solsona-Espriu, B.; Carley, A. F.; Herzing, A. A.; Watanabe, M.; Kiely, C. J.; Knight, D. W.; Hutchings, G. J., Solvent-free oxidation of primary alcohols to

aldehydes using Au-Pd/TiO2 catalysts. *Science* 2006, *311* (5759), 362-365.

[26] Miyamura, H.; Matsubara, R.; Miyazaki, Y.; Kobayashi, S., Aerobic oxidation of alcohols at room temperature and atmospheric conditions catalyzed by reusable gold nanoclusters stabilized by the benzene rings of polystyrene derivatives. *Angew Chem Int Edit* 2007, *46* (22), 4151-4154.

[27] Liu, Y. X.; Zhao, G. F.; Wang, D. S.; Li, Y. D., Heterogeneous catalysis for green chemistry based on nanocrystals. *Natl Sci Rev* 2015, *2* (2), 150-166.

[28] Wight, A. P.; Davis, M. E., Design and preparation of organic-inorganic hybrid catalysts. *Chem Rev* 2002, *102* (10), 3589-3613.

[29] Parmeggiani, C.; Matassini, C.; Cardona, F., A step forward towards sustainable aerobic alcohol oxidation: new and revised catalysts based on transition metals on solid supports. *Green Chem* 2017, *19* (9), 2030-2050.

[30] McMorn, P.; Hutchings, G. J., Heterogeneous enantioselective catalysts: strategies for the immobilisation of homogeneous catalysts. *Chem Soc Rev* 2004, *33* (2), 108-122.

[31] Li, C., Chiral synthesis on catalysts immobilized in microporous and mesoporous materials. *Catal Rev* 2004, *46* (3-4), 419-492.

[32] Xiao, B.; Yuan, Q. C., Nanoporous metal organic framework materials for hydrogen storage. *Particuology* 2009, *7* (2), 129-140.

[33] Liu, Y. D.; Kim, J.; Ahn, W. S.; Choi, H. J., Novel electrorheological properties of a metal-organic framework Cu-3(BTC)(2). *Chem Commun* 2012, *48* (45), 5635-5637.

[34] Kim, J.; Cho, H. Y.; Ahn, W. S., Synthesis and Adsorption/Catalytic Properties of the Metal Organic Framework CuBTC. *Catal Surv Asia* 2012, *16* (2), 106-119.

[35] Dhakshinamoorthy, A.; Alvaro, M.; Garcia, H., Aerobic Oxidation of Benzylic Alcohols Catalyzed by Metal-Organic Frameworks Assisted by TEMPO. *Acs Catal* 2011, *1* (1), 48-53.

[36] Peng, L.; Zhang, J. L.; Xue, Z. M.; Han, B. X.; Sang, X. X.; Liu, C. C.; Yang, G. Y., Highly mesoporous metal-organic framework assembled in a switchable solvent. *Nat Commun* 2014, *5*.

[37] Qi, Y.; Luan, Y.; Yu, J.; Peng, X.; Wang, G., Nanoscaled Copper Metal-Organic Framework (MOF) Based on Carboxylate Ligands as an Efficient Heterogeneous Catalyst for Aerobic Epoxidation of Olefins and Oxidation of Benzylic and Allylic Alcohols. *Chem-Eur J* 2015, *21* (4), 1589-1597.

[38] Chen, S. C.; Lu, S. N.; Tian, F.; Li, N.; Qian, H. Y.; Cui, A. J.; He, M. Y.; Chen, Q., Highly selective aerobic oxidation of alcohols to aldehydes over a new Cu(II)-based metal-organic framework with mixed linkers. *Catal Commun* 2017, *95*, 6-11.

[39] Nasani, R.; Saha, M.; Mobin, S. M.; Martins, L. M. D. R. S.; Pombeiro, A. J. L.; Kirillov, A. M.; Mukhopadhyay, S., Copper-organic frameworks assembled from in situ generated 5-(4-pyridyl)tetrazole building blocks: synthesis, structural features, topological analysis and catalytic oxidation of alcohols. *Dalton T* 2014, *43* (26), 9944-9954.

[40] Tu, B. B.; Pang, Q. Q.; Xu, H. S.; Li, X. M.; Wang, Y. L.; Ma, Z.; Weng, L. H.; Li, Q. W., Reversible Redox Activity in Multicomponent Metal-Organic Frameworks Constructed from Trinuclear Copper Pyrazolate Building Blocks. *J Am Chem Soc* 2017, *139* (23), 7998-8007.

[41] Kim, B. R.; Oh, J. S.; Kim, J.; Lee, C. Y., Aerobic Oxidation of Alcohols over Copper-containing Metal-Organic Frameworks. *B Korean Chem Soc* 2015, *36* (12), 2799-2800.

[42] Aryanejad, S.; Bagherzade, G.; Farrokhi, A., A nanoscale Cu-metal organic framework with Schiff base ligand: Synthesis, characterization and investigation catalytic activity in the oxidation of alcohols. *Inorg Chem Commun* 2017, *81*, 37-42.

[43] Karmakar, A.; Martins, L. M. D. R. S.; Hazra, S.; da Silva, M. F. C. G.; Pombeiro, A. J. L., Metal-Organic Frameworks with Pyridyl-Based Isophthalic Acid and Their Catalytic Applications in

Microwave Assisted Peroxidative Oxidation of Alcohols and Henry Reaction. *Cryst Growth Des* 2016, *16* (4), 1837-1849.

[44] Peng, L.; Wu, S. J.; Yang, X. Y.; Hu, J.; Fu, X. R.; Li, M.; Bai, L.; Huo, Q. S.; Guan, J. Q., Oxidation of benzyl alcohol over metal organic frameworks M-BTC (M = Co, Cu, Fe). *New J Chem* 2017, *41* (8), 2891-2894.

[45] Peng, L.; Wu, S. J.; Yang, X. Y.; Hu, J.; Fu, X. R.; Huo, Q. S.; Guan, J. Q., Application of metal organic frameworks M(bdc)(ted)(0.5) (M = Co, Zn, Ni, Cu) in the oxidation of benzyl alcohol. *Rsc Adv* 2016, *6* (76), 72433-72438.

[46] Zakzeski, J.; Debczak, A.; Bruijnincx, P. C. A.; Weckhuysen, B. M., Catalytic oxidation of aromatic oxygenates by the heterogeneous catalyst Co-ZIF-9. *Appl Catal a-Gen* 2011, *394* (1-2), 79-85.

[47] Farrokhi, A.; Jafarpour, M.; Najafzade, R., Phosphonate-based Metal Organic Frameworks as Robust Heterogeneous Catalysts for TBHP Oxidation of Benzylic Alcohols. *Catal Lett* 2017, *147* (7), 1714-1721.

[48] Wang, J. C.; Ding, F. W.; Ma, J. P.; Liu, Q. K.; Cheng, J. Y.; Dong, Y. B., Co(II)-MOF: A Highly Efficient Organic Oxidation Catalyst with Open Metal Sites. *Inorg Chem* 2015, *54* (22), 10865-10872.

[49] Mitchell, L.; Williamson, P.; Ehrlichova, B.; Anderson, A. E.; Seymour, V. R.; Ashbrook, S. E.; Acerbi, N.; Daniels, L. M.; Walton, R. I.; Clarke, M. L.; Wright, P. A., Mixed-Metal MIL-100(Sc,M) (M=Al, Cr, Fe) for Lewis Acid Catalysis and Tandem C-C Bond Formation and Alcohol Oxidation. *Chem-Eur J* 2014, *20* (51), 17185-17197.

[50] Rapeyko, A.; Arias, K. S.; Climent, M. J.; Corma, A.; Iborra, S., Polymers from biomass: one pot two-step synthesis of furilydenepropanenitrile derivatives with MIL-100(Fe) catalyst. *Catal Sci Technol* 2017, *7* (14), 3008-3016.

[51] Zhan, G. W.; Ng, W. C.; Lin, W. Y.; Koh, S. N.; Wang, C. H., Effective Recovery of Vanadium from Oil Refinery Waste into Vanadium-Based Metal-Organic Frameworks. *Environ Sci Technol* 2018, *52* (5), 3008-3015.

[52] Xamena, F. X. L. I.; Abad, A.; Corma, A.; Garcia, H., MOFs as catalysts: Activity, reusability and shape-selectivity of a Pd-containing MOF. *J Catal* 2007, *250* (2), 294-298.

[53] Lammert, M.; Wharmby, M. T.; Smolders, S.; Bueken, B.; Lieb, A.; Lomachenko, K. A.; De Vos, D.; Stock, N., Cerium-based metal organic frameworks with UiO-66 architecture: synthesis, properties and redox catalytic activity. *Chem Commun* 2015, *51* (63), 12578-12581.

[54] Smolders, S.; Lomachenko, K. A.; Bueken, B.; Struyf, A.; Bugaev, A. L.; Atzori, C.; Stock, N.; Lamberti, C.; Roeffaers, M. B. J.; De Vos, D. E., Unravelling the Redox-catalytic Behavior of Ce4+ Metal-Organic Frameworks by X-ray Absorption Spectroscopy. *Chemphyschem* 2018, *19* (4), 373-378.

[55] Sava Gallis, D. F.; Chapman, K. W.; Rodriguez, M. A.; Greathouse, J. A.; Parkes, M. V.; Nenoff', T. M., Selective O-2 Sorption at Ambient Temperatures via Node Distortions in Sc-MIL-100. *Chem Mater* 2016, *28* (10), 3327-3336.

[56] Wang, Z. Q.; Cohen, S. M., Postsynthetic covalent modification of a neutral metal-organic framework. *J Am Chem Soc* 2007, *129* (41), 12368-+.

[57] Tanabe, K. K.; Cohen, S. M., Postsynthetic modification of metal-organic frameworks-a progress report. *Chem Soc Rev* 2011, *40* (2), 498-519.

[58] Hou, J. Y.; Luan, Y.; Tang, J.; Wensley, A. M.; Yang, M.; Lu, Y. F., Synthesis of UiO-66-NH2 derived heterogeneous copper (II) catalyst and study of its application in the selective aerobic oxidation of alcohols. *J Mol Catal a-Chem* 2015, *407*, 53-59.

[59] Yuan, S.; Chen, Y. P.; Qin, J. S.; Lu, W. G.; Zou, L. F.; Zhang, Q.; Wang, X.; Sun, X.; Zhou, H. C., Linker Installation: Engineering Pore Environment with Precisely Placed Functionalities in Zirconium MOFs. *J Am Chem Soc* 2016, *138* (28), 8912-8919.

[60] Taher, A.; Kim, D. W.; Lee, I. M., Highly efficient metal organic framework (MOF)-based copper catalysts for the base-free aerobic oxidation of various alcohols. *Rsc Adv* 2017, *7* (29), 17806-17812.

[61] Thompson, A. B.; Pahis, D. R.; Bernales, V.; Gallington, L. C.; Malonzo, C. D.; Webber, T.; Tereniak, S. J.; Wang, T. C.; Desai, S. P.; Li, Z. Y.; Kim, I. S.; Gagliardi, L.; Penn, R. L.; Chapman, K. W.; Stein, A.; Farha, O. K.; Hupp, J. T.; Martinson, A. B. F.; Lu, C. C., Installing Heterobimetallic Cobalt-Aluminum Single Sites on a Metal Organic Framework Support. *Chem Mater* 2016, *28* (18), 6753-6762.

[62] Abednatanzi, S.; Derakhshandeh, P. G.; Abbasi, A.; Van der Voort, P.; Leus, K., Direct Synthesis of an Iridium(III) Bipyridine Metal-Organic Framework as a Heterogeneous Catalyst for Aerobic Alcohol Oxidation. *Chemcatchem* 2016, *8* (23), 3672-3679.

[63] Wu, S. T.; Chen, L. Y.; Yin, B. L.; Li, Y. W., "Click" post-functionalization of a metal-organic framework for engineering active single-site heterogeneous Ru(III) catalysts. *Chem Commun* 2015, *51* (48), 9884-9887.

[64] Liu, H.; Ramella, D.; Yu, P.; Luan, Y., A novel modified MIL-101-NH2 ligand for CuI-catalyzed and air promoted oxidation of secondary alcohols. *Rsc Adv* 2017, *7* (36), 22353-22359.

[65] Zalomaeva, O. V.; Kovalenko, K. A.; Chesalov, Y. A.; Mel'gunov, M. S.; Zaikovskii, V. I.; Kaichev, V. V.; Sorokin, A. B.; Kholdeeva, O. A.; Fedin, V. P., Iron tetrasulfophthalocyanine immobilized on metal organic framework MIL-101: synthesis, characterization and catalytic properties. *Dalton T* 2011, *40* (7), 1441-1444.

[66] Balu, A. M.; Lin, C. S. K.; Liu, H. L.; Li, Y. W.; Vargas, C.; Luque, R., Iron oxide functionalised MIL-101 materials in aqueous phase selective oxidations. *Appl Catal a-Gen* 2013, *455*, 261-266.

[67] Tarasov, A. L.; Kustov, L. M.; Isaeva, V. I.; Kalenchuk, A. N.; Mishin, I. V.; Kapustin, G. I.; Bogdan, V. I., Platinum-containing catalyst supported on a metal-organic framework structure in the selective oxidation of benzyl alcohol derivatives into aldehydes. *Kinet Catal+* 2011, *52* (2), 273-276.

[68] Carson, F.; Agrawal, S.; Gustafsson, M.; Bartoszewicz, A.; Moraga, F.; Zou, X. D.; Martin-Matute, B., Ruthenium Complexation in an Aluminium Metal-Organic Framework and Its Application in Alcohol Oxidation Catalysis. *Chem-Eur J* 2012, *18* (48), 15337-15344.

[69] Cui, Y. X.; Rimoldi, M.; Platero-Prats, A. E.; Chapman, K. W.; Hupp, J. T.; Farha, O. K., Stabilizing a Vanadium Oxide Catalyst by Supporting on a Metal-Organic Framework. *Chemcatchem* 2018, *10* (8), 1772-1777.

[70] Shu, X.; Yu, Y.; Jiang, Y.; Luan, Y.; Ramella, D., Direct synthesis of Fe(III) immobilized Zr-based metal-organic framework for aerobic oxidation reaction. *Appl Organomet Chem* 2017, *31* (12).

[71] Fei, H. H.; Shin, J. W.; Meng, Y. S.; Adelhardt, M.; Sutter, J.; Meyer, K.; Cohen, S. M., Reusable Oxidation Catalysis Using Metal-Monocatecholato Species in a Robust Metal-Organic Framework. *J Am Chem Soc* 2014, *136* (13), 4965-4973.

[72] Du, X.; Luan, Y.; Yang, F. X.; Ramella, D.; Shu, X., Picolinoyl functionalized MOF ligands for an air-promoted secondary alcohol oxidation with CuBr. *New J Chem* 2017, *41* (11), 4400-4405.

[73] Abednatanzi, S.; Leus, K.; Derakhshandeh, P. G.; Nahra, F.; De Keukeleere, K.; Van Hecke, K.; Van Driessche, I.; Abbasi, A.; Nolan, S. P.; Van Der Voort, P., POM@IL-MOFs - inclusion of POMs in ionic liquid modified MOFs to produce recyclable oxidation catalysts. *Catal Sci Technol* 2017, *7* (7), 1478-1487.

[74] Abednatanzi, S.; Abbasi, A.; Masteri-Farahani, M., Immobilization of catalytically active polyoxotungstate into ionic liquid-modified MIL-100(Fe): A recyclable catalyst for selective oxidation of benzyl alcohol. *Catal Commun* 2017, *96*, 6-10.

[75] Mizuno, N.; Misono, M., Heterogenous catalysis. *Chem Rev* 1998, *98* (1), 199-217.

[76] Mizuno, N.; Yamaguchi, K.; Kamata, K., Epoxidation of olefins with hydrogen peroxide catalyzed by polyoxometalates. *Coordin Chem Rev* 2005, *249* (17-18), 1944-1956.

[77] Maksimchuk, N. V.; Zalomaeva, O. V.; Skobelev, I. Y.; Kovalenko, K. A.; Fedin, V. P.; Kholdeeva, O. A., Metal-organic frameworks of the MIL-101 family as heterogeneous single-site catalysts. *P Roy Soc a-Math Phy* 2012, *468* (2143), 2017-2034.

[78] Luan, Y.; Qi, Y.; Gao, H. Y.; Zheng, N. N.; Wang, G., Synthesis of an amino-functionalized metal-organic framework at a nanoscale

level for gold nanoparticle deposition and catalysis. *J Mater Chem A* 2014, *2* (48), 20588-20596.

[79] Leus, K.; Concepcion, P.; Vandichel, M.; Meledina, M.; Grirrane, A.; Esquivel, D.; Turner, S.; Poelman, D.; Waroquier, M.; Van Speybroeck, V.; Van Tendeloo, G.; Garcia, H.; Van Der Voort, P., Au@UiO-66: a base free oxidation catalyst. *Rsc Adv* 2015, *5* (29), 22334-22342.

[80] Liu, H. L.; Liu, Y. L.; Li, Y. W.; Tang, Z. Y.; Jiang, H. F., Metal-Organic Framework Supported Gold Nanoparticles as a Highly Active Heterogeneous Catalyst for Aerobic Oxidation of Alcohols. *J Phys Chem C* 2010, *114* (31), 13362-13369.

[81] Zhu, J.; Wang, P. C.; Lu, M., Selective oxidation of benzyl alcohol under solvent-free condition with gold nanoparticles encapsulated in metal-organic framework. *Appl Catal a-Gen* 2014, *477*, 125-131.

[82] Liu, L. L.; Song, Y. B.; Chong, H. B.; Yang, S.; Xiang, J.; Jin, S.; Kang, X.; Zhang, J.; Yu, H. Z.; Zhu, M. Z., Size-confined growth of atom-precise nanoclusters in metal-organic frameworks and their catalytic applications. *Nanoscale* 2016, *8* (3), 1407-1412.

[83] Fu, Y.; Guo, Y. L.; Guo, Y.; Wang, Y. S.; Wang, L.; Zhan, W. C.; Lu, G. Z., In situ assembly of ultrafine Mn3O4 nanoparticles into MIL-101 for selective aerobic oxidation. *Catal Sci Technol* 2017, *7* (18), 4136-4144.

[84] Proch, S.; Herrmannsdorfer, J.; Kempe, R.; Kern, C.; Jess, A.; Seyfarth, L.; Senker, J., Pt@MOF-177: Synthesis, Room-Temperature Hydrogen Storage and Oxidation Catalysis. *Chem-Eur J* 2008, *14* (27), 8204-8212.

[85] Liu, H. L.; Chang, L. N.; Bai, C. H.; Chen, L. Y.; Luque, R.; Li, Y. W., Controllable Encapsulation of "Clean" Metal Clusters within MOFs through Kinetic Modulation: Towards Advanced Heterogeneous Nanocatalysts. *Angew Chem Int Edit* 2016, *55* (16), 5019-5023.

[86] Chen, L. Y.; Chen, H. R.; Luque, R.; Li, Y. W., Metal-organic framework encapsulated Pd nanoparticles: towards advanced heterogeneous catalysts. *Chem Sci* 2014, *5* (10), 3708-3714.

[87] Abedi, S.; Morsali, A., Improved activity of palladium nanoparticles using a sulfur-containing metal-organic framework as an efficient catalyst for selective aerobic oxidation in water. *New J Chem* 2017, *41* (13), 5846-5852.

[88] Akbari, S.; Mokhtari, J.; Mirjafari, Z., Solvent-free and melt aerobic oxidation of benzyl alcohols using Pd/Cu-2(BDC)(2)DABCO-MOF prepared by one-step and through reduction by dimethylformamide. *Rsc Adv* 2017, *7* (65), 40881-40886.

[89] Chen, G. Z.; Wu, S. J.; Liu, H. L.; Jiang, H. F.; Li, Y. W., Palladium supported on an acidic metal-organic framework as an efficient catalyst in selective aerobic oxidation of alcohols. *Green Chem* 2013, *15* (1), 230-235.

[90] Rosler, C.; Esken, D.; Wiktor, C.; Kobayashi, H.; Yamamoto, T.; Matsumura, S.; Kitagawa, H.; Fischer, R. A., Encapsulation of Bimetallic Nanoparticles into a Metal-Organic Framework: Preparation and Microstructure Characterization of Pd/Au@ZIF-8. *Eur J Inorg Chem* 2014, (32), 5514-5521.

[91] Bhattacharjee, S.; Kim, J.; Ahn, W. S., Pd Nanoparticles Supported on MIL-101: An Efficient Recyclable Catalyst in Oxidation and Hydrogenation Reactions. *J Nanosci Nanotechno* 2014, *14* (3), 2546-2552.

[92] Liu, H. L.; Chang, L. N.; Chen, L. Y.; Li, Y. W., In situ one-step synthesis of metal-organic framework encapsulated naked Pt nanoparticles without additional reductants. *J Mater Chem A* 2015, *3* (15), 8028-8033.

[93] Jiang, Z. W.; Dai, F. Q.; Huang, C. Z.; Li, Y. F., Facile synthesis of a Fe3O4/MIL-101(Fe) composite with enhanced catalytic performance. *Rsc Adv* 2016, *6* (89), 86443-86446.

[94] Kang, X. C.; Zhang, J. L.; Shang, W. T.; Wu, T. B.; Zhang, P.; Han, B. X.; Wu, Z. H.; Mo, G.; Xing, X. Q., One-Step Synthesis of Highly Efficient Nanocatalysts on the Supports with Hierarchical Pores Using Porous Ionic Liquid-Water Gel. *J Am Chem Soc* 2014, *136* (10), 3768-3771.

[95] Muller, M.; Turner, S.; Lebedev, O. I.; Wang, Y. M.; van Tendeloo, G.; Fischer, R. A., Au@MOF-5 and Au/MOx@MOF-5 (M = Zn, Ti; x=1, 2): Preparation and Microstructural Characterisation. *Eur J Inorg Chem* 2011, (12), 1876-1887.

[96] Ramezanalizadeh, H.; Manteghi, F., A novel MOF/BiFeO3 composite as a catalyst for efficient and selective oxidation of alcohols. *Rsc Adv* 2016, *6* (101), 99096-99104.

[97] Abedi, S.; Morsali, A., Ordered Mesoporous Metal Organic Frameworks Incorporated with Amorphous TiO2 As Photocatalyst for Selective Aerobic Oxidation in Sunlight Irradiation. *Acs Catal* 2014, *4* (5), 1398-1403.

[98] Naghdi, Z.; Farzaeli, R.; Aliyan, H., Building MOF bottles (MIL-101 family as heterogeneous single-site catalysts) around Fe3O4 ships: A highly efficient and magnetically separable catalyst for oxidation of alcohols. *Russ J Appl Chem+* 2015, *88* (8), 1343-1350.

[99] Saikia, M.; Bhuyan, D.; Saikia, L., Facile synthesis of Fe3O4 nanoparticles on metal organic framework MIL-101(Cr): characterization and catalytic activity. *New J Chem* 2015, *39* (1), 64-67.

[100] Li, J.; Gao, H. Y.; Tan, L.; Luan, Y.; Yang, M., Superparamagnetic Core-Shell Metal-Organic Framework Fe3O4/Cu-3(btc)(2) Microspheres and Their Catalytic Activity in the Aerobic Oxidation of Alcohols and Olefins. *Eur J Inorg Chem* 2016, (30), 4906-4912.

[101] Hou, J. Y.; Luan, Y.; Yu, J.; Qi, Y.; Wang, G.; Lu, Y. F., Fabrication of hierarchical composite microspheres of copper-doped Fe_3O_4@P4VP@ZIF-8 and their application in aerobic oxidationt. *New J Chem* 2016, *40* (12), 10127-10135.

[102] Miao, Z.; Shu, X.; Ramella, D., Synthesis of a Fe3O4@P4VP@metal-organic framework core-shell structure and studies of its aerobic oxidation reactivity. *Rsc Adv* 2017, *7* (5), 2773-2779.

[103] Xu, J.; He, S.; Zhang, H. L.; Huang, J. C.; Lin, H. X.; Wang, X. X.; Long, J. L., Layered metal-organic framework/graphene nanoarchitectures for organic photosynthesis under visible light. *J Mater Chem A* 2015, *3* (48), 24261-24271.

[104] Cai, J.; Lu, J. Y.; Chen, Q. Y.; Qu, L. L.; Lu, Y. Q.; Gao, G. F., Eu-Based MOF/graphene oxide composite: a novel photocatalyst for the oxidation of benzyl alcohol using water as oxygen source. *New J Chem* 2017, *41* (10), 3882-3886.

[105] Yang, X. Y.; Wu, S. J.; Hu, J.; Fu, X. R.; Peng, L.; Kan, Q. B.; Huo, Q. S.; Guan, J. Q., Highly efficient N-doped magnetic cobalt-graphene composite for selective oxidation of benzyl alcohol. *Catal Commun* 2016, *87*, 90-93.

[106] Desai, S. P.; Malonzo, C. D.; Webber, T.; Duan, J. X.; Thompson, A. B.; Tereniak, S. J.; DeStefano, M. R.; Buru, C. T.; Li, Z. Y.; Penn, R. L.; Farha, O. K.; Hupp, J. T.; Stein, A.; Lu, C. C., Assembly of dicobalt and cobalt-aluminum oxide clusters on metal-organic framework and nanocast silica supports. *Faraday Discuss* 2017, *201*, 299-314.

[107] Pascanu, V.; Gomez, A. B.; Ayats, C.; Platero-Prats, A. E.; Carson, F.; Su, J.; Yao, Q. X.; Pericas, M. A.; Zou, X. D.; Martin-Matute, B., Double-Supported Silica-Metal-Organic Framework Palladium Nanocatalyst for the Aerobic Oxidation of Alcohols under Batch and Continuous Flow Regimes. *Acs Catal* 2015, *5* (2), 472-479.

[108] Hashemian, S.; Sedrpoushan, A.; Eshbala, F. H., Co-Zeolite Imidazolate Frameworks (ZIF-9@Zeolite) as Heterogen Catalyst for Alcohols Oxidation. *Catal Lett* 2017, *147* (1), 196-203.

[109] Shen, L. J.; Liang, S. J.; Wu, W. M.; Liang, R. W.; Wu, L., CdS-decorated UiO-66(NH2) nanocomposites fabricated by a facile photodeposition process: an efficient and stable visible-light-driven photocatalyst for selective oxidation of alcohols. *J Mater Chem A* 2013, *1* (37), 11473-11482.

[110] Xue, Z. M.; Zhang, J. L.; Peng, L.; Han, B. X.; Mu, T. C.; Li, J. S.; Yang, G. Y., Poly(ethylene glycol) Stabilized Mesoporous Metal-Organic Framework Nanocrystals: Efficient and Durable Catalysts for the Oxidation of Benzyl Alcohol. *Chemphyschem* 2014, *15* (1), 85-89.

[111] Ge, C. H.; Du, Y.; Wang, R.; Xue, L. L.; Wu, Z. S.; Xing, T. Z.; Ji, X. C.; Ma, L.; Zhang, X. D., A facile strategy to fabricate carboxyl-rich

carbon spheres with copper-based MOFs through coordination bond. *J Porous Mat* 2016, *23* (6), 1537-1545.

[112] Kim, B. R.; Oh, J. S.; Kim, J.; Lee, C. Y., Robust Aerobic Alcohol Oxidation Catalyst Derived from Metal-Organic Frameworks. *Catal Lett* 2016, *146* (4), 734-743.

[113] Zhou, Y. X.; Chen, Y. Z.; Cao, L. N.; Lu, J. L.; Jiang, H. L., Conversion of a metal-organic framework to N-doped porous carbon incorporating Co and CoO nanoparticles: direct oxidation of alcohols to esters. *Chem Commun* 2015, *51* (39), 8292-8295.

[114] Zhong, W.; Liu, H. L.; Bai, C. H.; Liao, S. J.; Li, Y. W., Base-Free Oxidation of Alcohols to Esters at Room Temperature and Atmospheric Conditions using Nanoscale Co-Based Catalysts. *Acs Catal* 2015, *5* (3), 1850-1856.

[115] Chen, H. R.; Shen, K.; Mao, Q.; Chen, J. Y.; Li, Y. W., Nanoreactor of MOF-Derived Yolk-Shell Co@C-N: Precisely Controllable Structure and Enhanced Catalytic Activity. *Acs Catal* 2018, *8* (2), 1417-1426.

[116] Goh, T. W.; Xiao, C. X.; Maligal-Ganesh, R. V.; Li, X. L.; Huang, W. Y., Utilizing mixed-linker zirconium based metal-organic frameworks to enhance the visible light photocatalytic oxidation of alcohol. *Chem Eng Sci* 2015, *124*, 45-51.

[117] Chambers, M. B.; Wang, X.; Ellezam, L.; Ersen, O.; Fontecave, M.; Sanchez, C.; Rozes, L.; Mellot-Draznieks, C., Maximizing the Photocatalytic Activity of Metal-Organic Frameworks with Aminated-Functionalized Linkers: Substoichiometric Effects in MIL-125-NH2. *J Am Chem Soc* 2017, *139* (24), 8222-8228.

[118] Chen, Y. Z.; Wang, Z. U.; Wang, H. W.; Lu, J. L.; Yu, S. H.; Jiang, H. L., Singlet Oxygen-Engaged Selective Photo-Oxidation over Pt Nanocrystals/Porphyrinic MOF: The Roles of Photothermal Effect and Pt Electronic State. *J Am Chem Soc* 2017, *139* (5), 2035-2044.

[119] Yuan, S.; Liu, T. F.; Feng, D. W.; Tian, J.; Wang, K. C.; Qin, J. S.; Zhang, Q.; Chen, Y. P.; Bosch, M.; Zou, L. F.; Teat, S. J.; Dalgarno, S. J.; Zhou, H. C., A single crystalline porphyrinic titanium metal-organic framework. *Chem Sci* 2015, *6* (7), 3926-3930.

[120] Lian, X.; Yan, B., A Postsynthetic Modified MOF Hybrid as Heterogeneous Photocatalyst for alpha-Phenethyl Alcohol and Reusable Fluorescence Sensor. *Inorg Chem* 2016, *55* (22), 11831-11838.

[121] Chen, L. Y.; Peng, Y.; Wang, H.; Gua, Z. Z.; Duana, C. Y., Synthesis of Au@ZIF-8 single- or multi-core-shell structures for photocatalysis. *Chem Commun* 2014, *50* (63), 8651-8654.

[122] Shen, L. J.; Liang, S. J.; Wu, W. M.; Liang, R. W.; Wu, L., Multifunctional NH2-mediated zirconium metal-organic framework as an efficient visible-light-driven photocatalyst for selective oxidation of alcohols and reduction of aqueous Cr(VI). *Dalton T* 2013, *42* (37), 13649-13657.

[123] Ke, F.; Wang, L. H.; Zhu, J. F., Facile fabrication of CdS-metal-organic framework nanocomposites with enhanced visible-light photocatalytic activity for organic transformation. *Nano Res* 2015, *8* (6), 1834-1846.

[124] Wu, Z. Y.; Huang, X. B.; Zheng, H. Y.; Wang, P.; Hai, G. T.; Dong, W. J.; Wang, G., Aromatic heterocycle-grafted NH2-MIL-125(Ti) via conjugated linker with enhanced photocatalytic activity for selective oxidation of alcohols under visible light. *Appl Catal B-Environ* 2018, *224*, 479-487.

[125] Xu, X. Q.; Liu, R. X.; Cui, Y. H.; Liang, X. X.; Lei, C.; Meng, S. Y.; Ma, Y. L.; Lei, Z. Q.; Yang, Z. W., PANI/FeUiO-66 nanohybrids with enhanced visible-light promoted photocatalytic activity for the selectively aerobic oxidation of aromatic alcohols. *Appl Catal B-Environ* 2017, *210*, 484-494.

[126] Fu, Y. H.; Sun, L.; Yang, H.; Xu, L.; Zhang, F. M.; Zhu, W. D., Visible-light-induced aerobic photocatalytic oxidation of aromatic alcohols to aldehydes over Ni-doped NH2-MIL-125(Ti). *Appl Catal B-Environ* 2016, *187*, 212-217.

[127] Cote, A. P.; Benin, A. I.; Ockwig, N. W.; O'Keeffe, M.; Matzger, A. J.; Yaghi, O. M., Porous, crystalline, covalent organic frameworks. *Science* 2005, *310* (5751), 1166-1170.

[128] Kuhn, P.; Antonietti, M.; Thomas, A., Porous, covalent triazine-based frameworks prepared by ionothermal synthesis. *Angew Chem Int Edit* 2008, *47* (18), 3450-3453.

[129] Wang, K. W.; Yang, L. M.; Wang, X.; Guo, L. P.; Cheng, G.; Zhang, C.; Jin, S. B.; Tan, B.; Cooper, A., Covalent Triazine Frameworks via a Low-Temperature Polycondensation Approach. *Angew Chem Int Edit* 2017, *56* (45), 14149-14153.

[130] Meier, C. B.; Sprick, R. S.; Monti, A.; Guiglion, P.; Lee, J. S. M.; Zwijnenburg, M. A.; Cooper, A. I., Structure-property relationships for covalent triazine-based frameworks: The effect of spacer length on photocatalytic hydrogen evolution from water. *Polymer* 2017, *126*, 283-290.

[131] Uribe-Romo, F. J.; Hunt, J. R.; Furukawa, H.; Klock, C.; O'Keeffe, M.; Yaghi, O. M., A Crystalline Imine-Linked 3-D Porous Covalent Organic Framework. *J Am Chem Soc* 2009, *131* (13), 4570-+.

[132] Ding, S. Y.; Gao, J.; Wang, Q.; Zhang, Y.; Song, W. G.; Su, C. Y.; Wang, W., Construction of Covalent Organic Framework for Catalysis: Pd/COF-LZU1 in Suzuki-Miyaura Coupling Reaction. *J Am Chem Soc* 2011, *133* (49), 19816-19822.

[133] Ding, S. Y.; Wang, W., Covalent organic frameworks (COFs): from design to applications. *Chem Soc Rev* 2013, *42* (2), 548-568.

[134] Chandra, S.; Kandambeth, S.; Biswal, B. P.; Lukose, B.; Kunjir, S. M.; Chaudhary, M.; Babarao, R.; Heine, T.; Banerjee, R., Chemically Stable Multilayered Covalent Organic Nanosheets from Covalent Organic Frameworks via Mechanical Delamination. *J Am Chem Soc* 2013, *135* (47), 17853-17861.

[135] Kandambeth, S.; Mallick, A.; Lukose, B.; Mane, M. V.; Heine, T.; Banerjee, R., Construction of Crystalline 2D Covalent Organic Frameworks with Remarkable Chemical (Acid/Base) Stability via a Combined Reversible and Irreversible Route. *J Am Chem Soc* 2012, *134* (48), 19524-19527.

[136] Xu, H.; Gao, J.; Jiang, D. L., Stable, crystalline, porous, covalent organic frameworks as a platform for chiral organocatalysts. *Nat Chem* 2015, *7* (11), 905-912.

[137] Banerjee, T.; Gottschling, K.; Savasci, G.; Ochsenfeld, C.; Lotsch, B. V., H-2 Evolution with Covalent Organic Framework Photocatalysts. *Acs Energy Lett* 2018, *3* (2), 400-409.

[138] Huang, W.; Ma, B. C.; Lu, H.; Li, R.; Wang, L.; Landfester, K.; Zhang, K. A. I., Visible-Light-Promoted Selective Oxidation of Alcohols Using a Covalent Triazine Framework. *Acs Catal* 2017, *7* (8), 5438-5442.

[139] Khatioda, R.; Talukdar, D.; Saikia, B.; Bania, K. K.; Sarma, B., Constructing two dimensional amide porous polymer to promote selective oxidation reactions. *Catal Sci Technol* 2017, *7* (14), 3143-3150.

[140] Chan-Thaw, C. E.; Villa, A.; Prati, L.; Thomas, A., Triazine-Based Polymers as Nanostructured Supports for the Liquid-Phase Oxidation of Alcohols. *Chem-Eur J* 2011, *17* (3), 1052-1057.

[141] Chen, L. Y.; Zhang, L.; Chen, Z. J.; Liu, H. J.; Luque, R.; Li, Y. W., A covalent organic framework-based route to the in situ encapsulation of metal nanoparticles in N-rich hollow carbon spheres. *Chem Sci* 2016, *7* (9), 6015-6020.

[142] Zhou, Y.; Xiang, Z. H.; Cao, D. P.; Liu, C. J., Preparation and Characterization of Covalent Organic Polymer Supported Palladium Catalysts for Oxidation of CO and Benzyl Alcohol. *Ind Eng Chem Res* 2014, *53* (4), 1359-1367.

[143] Ishida, T.; Onuma, Y.; Kinjo, K.; Hamasaki, A.; Ohashi, H.; Honma, T.; Akita, T.; Yokoyama, T.; Tokunaga, M.; Haruta, M., Preparation of microporous polymer-encapsulated Pd nanoparticles and their catalytic performance for hydrogenation and oxidation. *Tetrahedron* 2014, *70* (36), 6150-6155.

[144] Yamaguchi, S.; Kamiya, K.; Hashimoto, K.; Nakanishi, S., Ru atom-modified covalent triazine framework as a robust electrocatalyst for selective alcohol oxidation in aqueous electrolytes. *Chem Commun* 2017, *53* (75), 10437-10440.

In: Alcohol Oxidation
Editor: Steffan Paterson

ISBN: 978-1-53614-604-2

Chapter 4

OXIDATIVE NON-ENZYMATIC KINETIC RESOLUTION OF RACEMIC ALCOHOLS AND RELATED COMPOUNDS

Maximilian N. Kopylovich** and Ana Maria Faisca Phillips
Centro de Química Estrutural, Instituto Superior Técnico,
Universidade de Lisboa, Lisbon, Portugal

ABSTRACT

Enantiopure alcohols are important starting materials and intermediates for the synthesis of many biologically active compounds and fine chemicals. One of the simplest and most convenient ways to access them is an oxidative kinetic resolution of racemic alcohols. The use of enzymes for the kinetic resolution of racemates has given affordable enantiopure compounds in good yields, however, non-enzymatic chiral catalytic systems have been gaining ground. Accordingly, the effective non-enzymatic oxidative kinetic resolutions of racemic alcohols and

---

* Corresponding Author Email: maximilian.kopylovich@tecnico.ulisboa.pt.

related compounds such as diols or benzoins (α-hydroxy ketones) are summarized and discussed in this chapter.

Keywords: oxidative kinetic resolution, enantiopure alcohols, enantiopure diols, enantiopure benzoins (α-hydroxy ketones)

1. Introduction

The majority of molecules in living organisms are chiral, and molecular interactions of the biomolecules will be different for different enantiomers of the same compound. As a result, both the US Food and Drug Administration and European Medicines Agency have demanded that the physiological and chemical properties of each enantiomer of a pharmaceutical product be individually submitted for evaluation. In addition to higher safety, lower dosages and improved efficacy, the extension of the patent term is also an incentive for the pharmaceutical companies to look for a single enantiomer. With the prospect that 95% of all the potentially racemic drugs will be chiral by 2030, synthesis of enantiopure compounds becomes crucial for development of new bioactive molecules. Enantiomerically pure compounds are also very sought after in the agrochemical and food industries. The reasons are similar to those for the pharmaceutical sector: the targeted synthesis of the respective active enantiomer leads to reduced applied quantities and thus improves the safety, economics and environmental impact. Continuing demand for the chirally pure compounds has pushed the development of new methodologies including synthesis from the naturally available chiral molecules, asymmetric synthesis from prochiral substrates and resolution of racemic mixtures [1–5].

Racemic secondary alcohols and related compounds, e.g., diols or benzoins (α-hydroxy ketones), are among the most abundant and cheap starting materials, thus, their deracemization can provide a significant added value to a number of chemical processes. One of the most straightforward deracemization methodologies is the oxidative kinetic resolution (OKR,

Scheme 1), where one of the enantiomers of a racemic mixture is more readily oxidized into e.g., ketone than its mirror image [1–5]. One should note that ketones are important intermediates by themselves and are widely used for further processing [6, 7]. Hence, OKR of the racemic *sec*-alcohols and related compounds allows to produce, in one step, two products with added value.

After the process, both the products (enantiopure *sec*-alcohol and ketone) can be easily separated due to the different physicochemical properties [6, 7] and used as intermetiates for further synthesis. Moreover, the formed prochiral ketone can be hydrogenated *in situ* to the racemic *sec*-alcohol and recycled. Typically, five cycles are required to produce the desired enantiomer with high yield (95%) [5]. In such a way, all the racemate can be converted into the respective enantiopure alcohol. In spite of the great potential, kinetic resolutions which involve further treatment and racemization (e.g., dynamic kinetic resolutions etc.) will not be discussed in detail along this chapter since this topic is well reviewed elsewhere [8]. Hence, here we will concentrate only on the direct one-step OKR of *sec*-alcohols and related compounds such as diols.

Enzymatic OKR protocols have been known for a long time, and significant progress in this field has been achieved and thoroughly reviewed [4, 5]. However, enzymatic processes are usually rather sensitive, expensive and difficult to scale-up; hence, a lot of effort has been made to elaborate practical artificial catalysts. Accordingly, fully non-enzymatic systems for the OKR of racemic *sec*-alcohols and prochiral diols have been recently proposed [1–3]. Apart from the mentioned above reference sources [1–5], there are several other related books, reviews and book chapters where the known information on this topic is fragmentally portrayed [9–15]. Accordinly, this chapter aims to generalize the knowledge on the enantiodifferentiating oxidations of racemic alcohols performed in the presence of chiral catalysts, with some notorious and recent examples discussed in more detail. The Chapter is divided into sections according to the nature of the catalysts used to promote the OKR reactions.

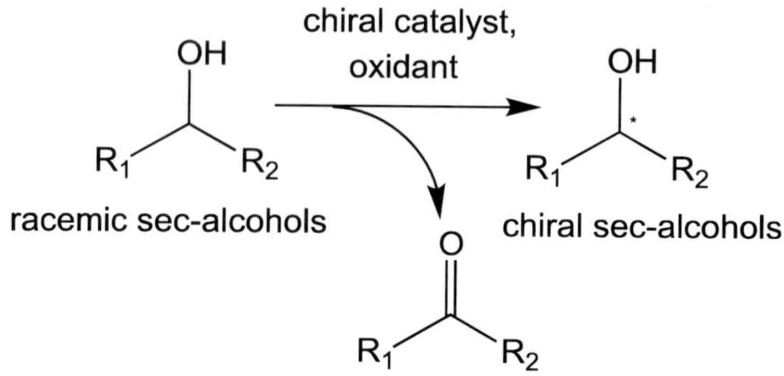

Scheme 1. Oxidative kinetic resolution (OKR) of secondary alcohols.

2. Organocatalytic Oxidations

In 1979 it was demonstrated [16] that chiral aminyl oxides react with benzoin to give benzil together with unreacted benzoin with an enantiomeric excess of *ca.* 7%. This study revealed that the found enantiomeric excess can potentially be amplified through several catalytic cycles if the related chiral nitroxides are applied as *catalytic* species. Accordingly, some of the first catalysts for the OKR of racemic secondary alcohols were nitroxyl radical **1** [17], chiral 1-azaspiro[5.5]undecane-*N*-oxyl **2** [18] or a combination of TEMPO (2,2,6,6-tetramethylpiperidin-1-yloxyl) **3** with chiral (−)-sparteine co-catalyst **4** [19] (Scheme 2). These catalysts were not quite adequate for the efficient enantioselective oxidation of unactivated alcohols, most probably due to their steric bulkiness. Accordingly, less hindered catalysts with enhanced catalytic activity and enantioselectivity (e.g., **5**, **6** and **7**) were then elaborated [20–28].

Later on, a number of AZADO-based (AZADO = 2-azaadamantane *N*-oxyl) chiral alkoxyamines were tested as precatalysts in OKR with trichloroisocyanuric acid (TCCA) as the co-oxidant in the presence of $NaHCO_3$ in CH_2Cl_2 solution at −40 °C, and the structure−enantioselectivity correlation study allowed the identification of 7-benzyl-3-*n*-butyl-4-oxa-5-azahomoadamantane (**6**, Scheme 3) as the most active and selective (pre)catalyst with the *ee* of the resulting alcohols reaching 99% [21, 22, 29].

The catalytic system was applicable to various unactivated aliphatic and heterocyclic secondary alcohols, while its synthetic potential was demonstrated by the synthesis of bioactive compounds of high enantiopurity [30]. It was also disclosed that the chlorine atom is crucial for high reactivity and enantioselectivity, since the chlorine-oxoammonium molecules were identified as active catalytic species generated *in situ* from the alkoxyamines (Scheme 3) [29].

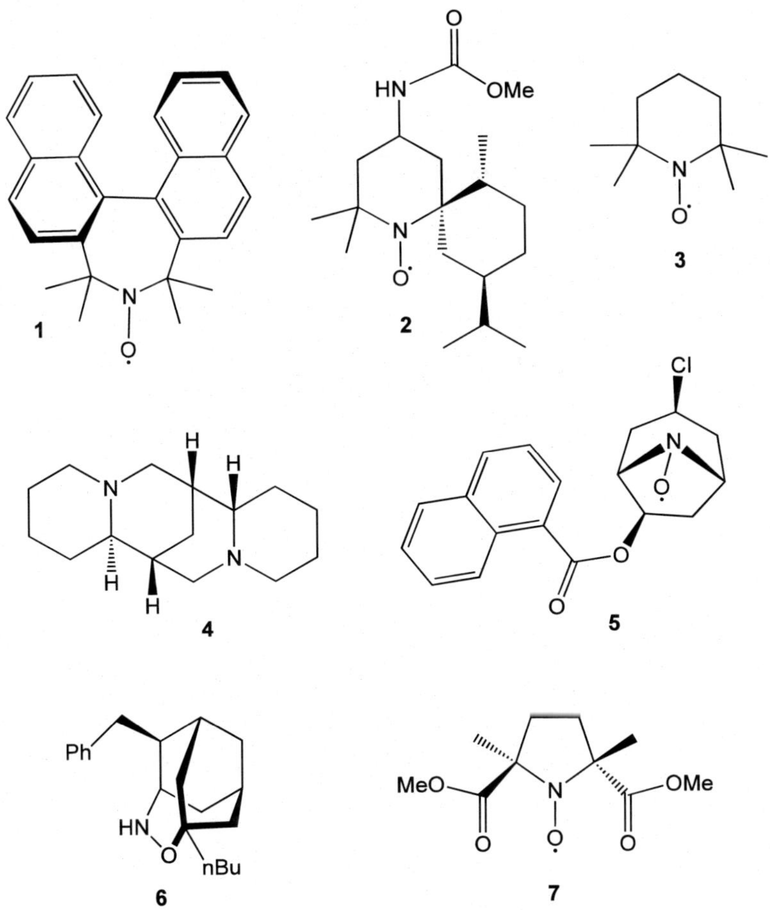

Scheme 2. Some nitroxyl-based catalysts.

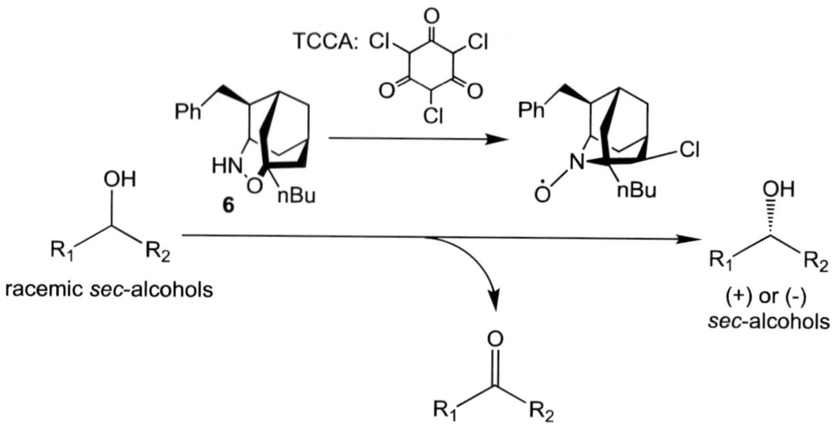

Scheme 3. OKR of various alcohols trough the generation of chiral oxoammonium ions from alkoxyamine **6** and TCCA.

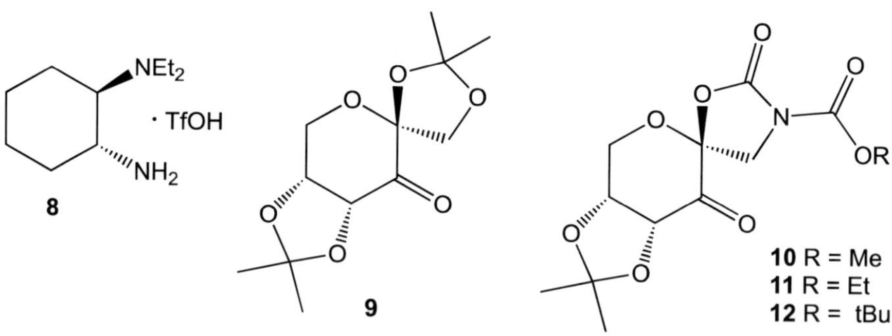

Scheme 4. Other types of organocatalysts for the OKR of *sec*-alcohols.

Along with the nitroxyl-based systems, some other types of organocatalysts for the OKR of racemic alcohols, *vic*-diols and *meso*-diols have been reported [31–37]. For instance, retro- and transfer-aldol reactions can be organocatalyzed by the chiral primary-tertiary diamine **8** (Scheme 4) with excellent enantio- (91–99% *ee*) and diastereoselectivities (>95:5) [33]. Other effective catalytic systems are based on Shi's fructose-derived ketone **9** and related oxazolidinone dioxiranes such as **10–12** (Scheme 4) [34–37].

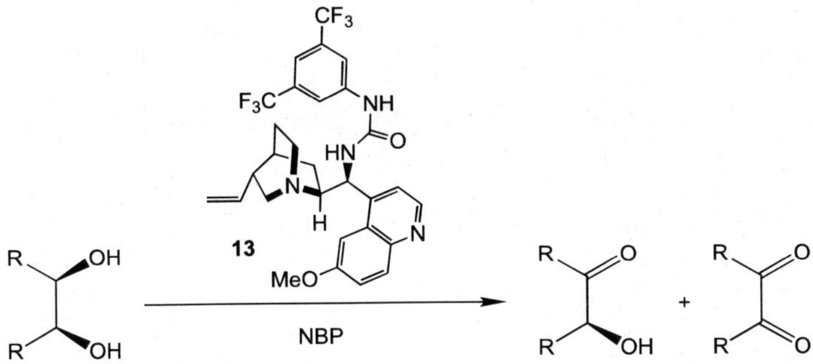

Scheme 5. Enantioselective oxidative desymmetrization of *meso*-1,2-diols catalyzed by the quinine-derived urea **13**.

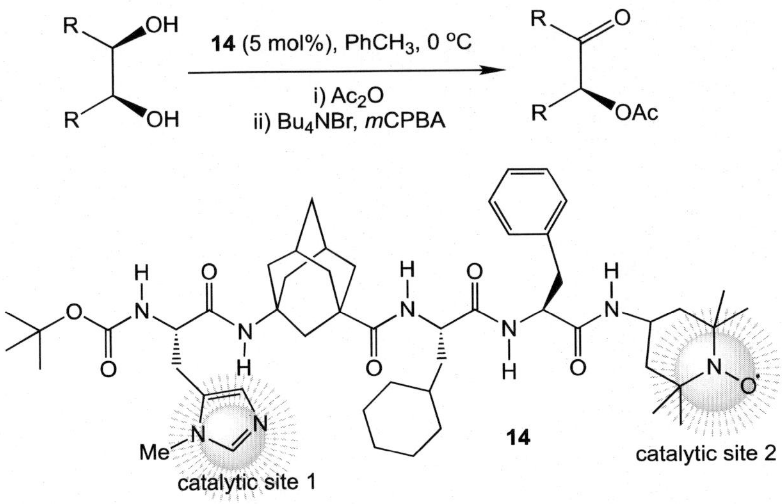

Scheme 6. Concurrent tandem desymmetrization/oxidation of meso-1,2-alkane diols with multicatalyst 14.

Quinine-derived ureas such as **13** (Scheme 5) are another type of the organocatalysts used for the enantioselective oxidation of a wide range of diaryl-substituted *meso*-1,2-diols using bromination reagents (e.g., *N*-bromophthalimide, NBP) as the oxidant [32]. Different *para*-, *meta*- and *ortho*-substitutions on the aryl unit can be tolerated, while the procedure operates at ambient temperature giving α-hydroxy ketones in yields as high as 94% and up to 95% *ee*. However, the catalytic system is not so efficient for dialkyl substituted-diols (yields up to 50% with 54% *ee*). It was proposed that H-bonding interactions between the urea moiety on the catalyst and the phthalimide and diol (serving as H-bond donor) are key factors that account for the enantioselectivity of this catalytic system.

The multicatalytic reactions in one pot can lead to a pronounced synergistic effect [6, 7, 38], and hence combination of the known catalytic motifs in one molecule can be a viable approach to improve yields and selectivities of the OKR reactions. One of the pioneering works in this direction concerns a multicatalyst system, where several orthogonal *(i.e.*, independent) organocatalytic moieties are combined together on an oligopeptide backbone (Scheme 6) [39]. The orthogonality in reactivity is achieved through the application of bulky and inert admantane spacer, which keeps the catalytic moieties apart and prevents folding of the backbone. This creates a one-pot multicatalyst "assembly line" with many operational advantages (e.g., decreased solvent use, time, and simpler workup) over traditional multistep synthesis [6, 7, 38]. Accordingly, catalytic acetylation and subsequent catalytic oxidation of *meso*-1,2-cycloalkane diols with *meta*-chloroperbenzoic acid (*m*CPBA) as an oxidant can be conducted in one-pot with good overall yields and *ee* values (Scheme 6). It was demonstrated that the incorporation of a TEMPO moiety in an oligopeptide sequence increases the oxidation activity compared to using TEMPO itself, either in the same sequence of reactions or in a separate reaction [39].

3. Transition-Metal Catalysts with Chiral Ligands

Along with the above-mentioned organocatalytic systems, several effective metal-based non-enzymatic catalysts for the OKR of racemic alcohols and related compounds have been also developed and reviewed [1-15, 40]. One of the first catalysts of this type was reported in 1997 and consists of a chiral diamine *ruthenium* catalyst **15** (Scheme 7) and acetone as the hydride acceptor [41]. Later on several chiral ferrocenyloxazolinylphosphine [42] and arsinooxazoline [43] ligands have been elaborated and combined with ruthenium(II) to provide the OKR of 1-arylalkanols. Moreover, chiral ruthenium–salen complexes (e.g., **16–20**, Scheme 7) have also demonstrated their potential as versatile and effective catalysts (enantioselectivities of up to 99% *ee*) for the photoinduced OKR of activated and unactivated aromatic and aliphatic secondary alcohols and diols with various oxidants including air oxygen [44-50].

Using oxygen from air as the primary oxidant for the OKR of alcohols in the absence of photoirradiation is very attractive from a green point of view, and this was first realized in 2001 with *palladium* catalysts (e.g., **21–23**, Scheme 8) in the presence of naturally occurring diamine (−)-sparteine as a chiral ligand [51]. The application of such catalytic systems has permitted the successful resolution of a number of benzylic, allylic and cyclopropyl secondary alcohols and diols with high selectivity [9, 10, 51–60]. Moreover, this methodology constituted some key steps in the syntheses of several alkaloids and pharmaceuticals [60–64]. Other sparteine analogues, such as (−)-*R*-isosparteine, (+)-α-isosparteine and chiral 2-*endo*- substituted 9- oxabispidines, have been ealaborated and applied for the OKR of alkohols in the following years [9, 10, 65–68]. Whilst the use of Pd-sparteine complexes is rather effective, however, this ligand is expensive and sensitive. Accordingly, alternative axially chiral Pd(II) complexes with *N*-heterocyclic carbene ligands derived from BINAM, e.g., **24–26** (Scheme 8), were developed later and successfully exploited [9, 10, 69, 70].

In 2003 several chiral salen *manganese* complexes (e.g.**, 27–29**, Scheme 9) have been successfully applied to achieve OKR of *sec*-alcohols in water/organic media with diacetoxyiodobenzene $PhI(OAc)_2$ as an oxidant [14, 71–74]. The line was then developed by several groups, and a number of related chiral monomeric, oligomeric, macrocyclic, polymeric, immobilized and supported salen manganese complexes have been used to achive the OKR of aliphatic, aromatic, aryl-alkyl or unsaturated secondary alcohols [14, 75–80]. The usage of the polymeric, immobilized and supported salen manganese complexes allowed the recovery of the catalyst and its reuse several times with retention of high activity and enantioselectivity [14, 75–80]. The spectrum of the appropriate oxidants has been also expanded, and *N*-bromosuccinimide (NBS), sodium hypochlorite and hydrogen peroxide were shown to be effective with the Mn(III)–salen complexes [81–83]. The mechanism of the stereochemical oxidation with the Mn(III)–salen complexes and NBS oxidant has been thoroughly studied [84], and it was proposed that it includes the formation of a Mn(V)-salen dibromide species with its subsequent reaction with alcohol to give an alkoxy-Mn(V) species. Then the carbonyl-forming elimination produces the ketone via a transition state with intramolecular hydrogen transfer from carbon phenoxy of the salen ligand [84]. Finally, it was demonstrated that the bioinspired chiral Mn(II) aminopyridine complexes (some examples are depicted on Scheme 9) can be also applied and provide asymmetric autoamplification in the OKR of secondary benzylic alcohols [85, 86].

The first report on the OKR of secondary arylalkanols with *iridium* catalysts involves catalytic hydrogen transfer between alcohols and ketones mediated by the chiral diaminodiphosphine PNNP ligand **33** (Scheme 10),

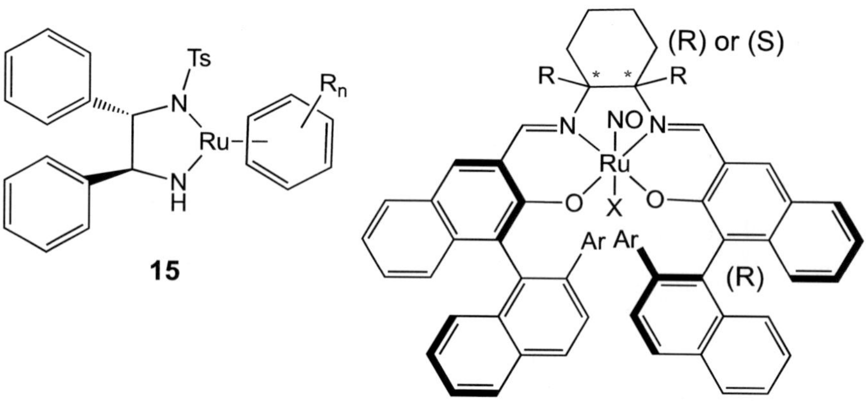

16: (S,R), R=H, Ar=Ph, X=Cl
17: (S,R), R=Me, Ar=Ph, X=Cl
18: (S,R), R=Me, Ar=4-PhC_6H_4, X=OH
19: (R,R), R=H, Ar=Ph, X=Cl
20: (R,R), R=Me, Ar=Ph, X=Cl

Scheme 7. Examples of Ru-based catalysts for the OKR of racemic alcohols.

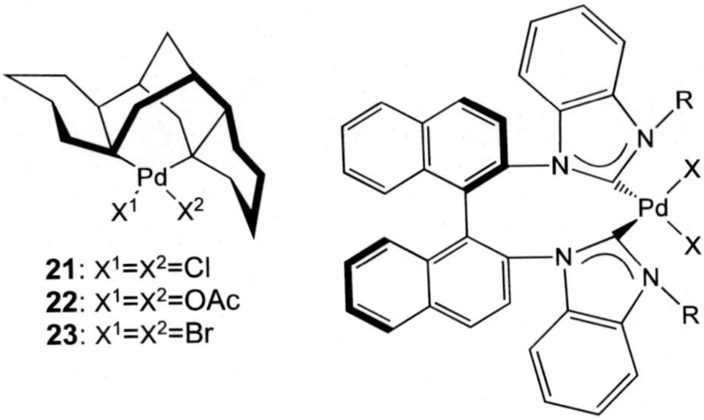

21: $X^1=X^2=Cl$
22: $X^1=X^2=OAc$
23: $X^1=X^2=Br$

24: R=Me, X=I
25: R=Et, X=I
26: R=Bn, X=I

Scheme 8. Representatives of Pd-based catalysts.

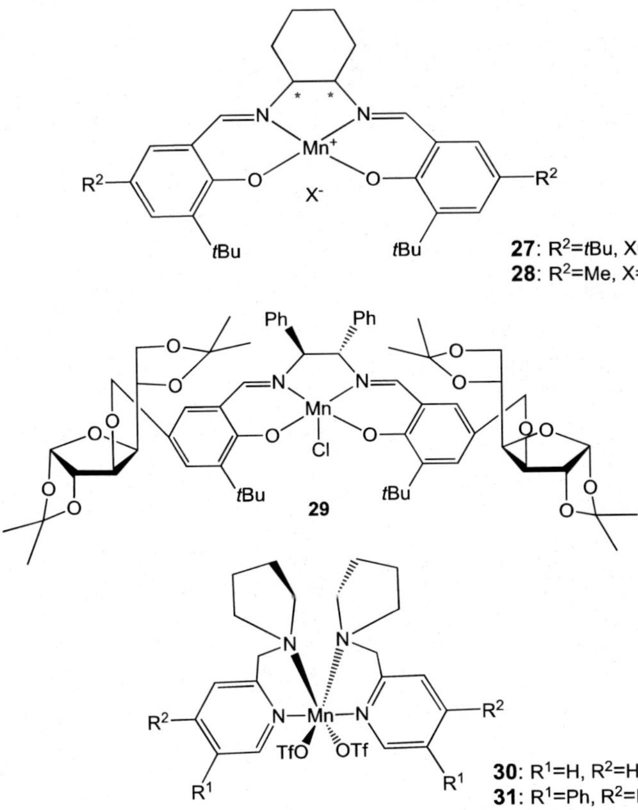

27: R^2=*t*Bu, X=Cl
28: R^2=Me, X=PF_6

29

30: R^1=H, R^2=H
31: R^1=Ph, R^2=H
32: R^1=H, R^2=OCH_2CF_3

Scheme 9. Examples of the chiral Mn(III) salen (**27–29**) and aminopyridine (**30–32**) complexes.

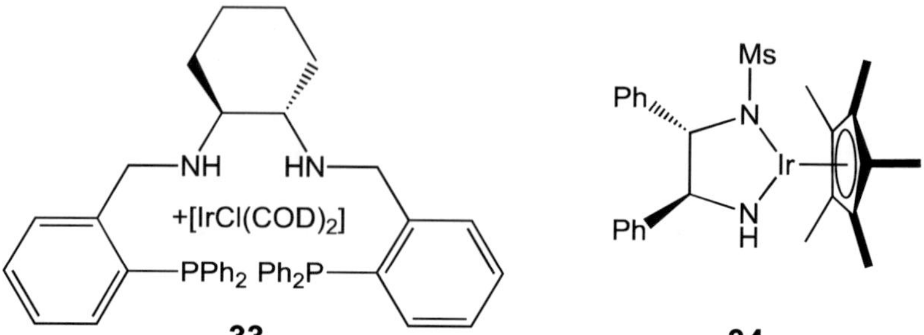

Scheme 10. Some Ir catalysts (COD = 1,5-cyclooctadiene).

and excellent enantioselectivities of 98% *ee* were achieved under mild conditions and in acetone as a dual role solvent/hydrogen acceptor [11, 87, 88]. Another effective (up to 99% *ee*) aerobic oxidative system employs half-sandwich iridium complex **34** bearing an *N*-sulfonylated diamine ligand (Scheme 10) [11, 89]. The scope of the latter catalyst **34** was than extended to a number of prochiral diols to give the corresponding chiral hydroxyl ketones [90, 91]. With the transformation of cyclic diols to the asymmetric lactones as a key step, a short-step synthesis of several bioactive molecules such as ottelione and scyphostatin has been developed [90].

Several other reported catalytic systems for the OKR of alcohols involve *vanadium* complexes, mainly based on the combination of $VO(O\textit{i}Pr)_3$ and chiral Schiff base ligands derived from 3,5-di-*tert*-butylsalicylaldehyde and (*S*)-*tert*-leucinol (*tert*-leucine) or chiral vanadyl carboxylates derived from *N*-salicylidene-L-α-amino acids and vanadyl sulfate (Scheme 11) [92–95]. Such catalysts allowed the aerobic and mild oxidation of a series of α-aryl- and α-heteroaryl-α-hydroxyphosphonates, α-hydroxyl esters and amides with diverse α-, O-, and N-substituents [92–98]. The developed protocols can be applied to racemic taxol C-13 side-chain resolution, OKR of methyl *o*-chloromandelate and synthesis of octalactin A and (–)-pantofuranoid E, demonstrating its great potential for the pharmaceutical industry [96–98]. The analogous **35–37** catalysts were supported on poly(styrene) resin or mesoporous silica, thus allowing the catalyst recovery and multiple recycle [99].

The successful application of inexpensive and environmentally benign *iron* for the OKR of racemic benzoins was first demonstrated in 2008 [100]. In this case (*R*)-BINAM-based ligand [(*R*)-BINAM = (*R*)-(+)-1,1′-binaphthyl-2,2′-diamine] was used in combination with Fe(II) acetate (**38**, Scheme 12), TEMPO co-catalyst and molecular oxygen as a terminal oxidant to achieve enantioselectivities of up to 98% *ee* [100]. The direction was further developed and generalized towards various benzylic alcohols, cyclic and alkenyl carbinols using iron(salan)(2-naphthoxo) catalyst **39** (Scheme 12) [101]. Other iron-based catalytic systems, e.g., with *trans*-(*R,R*)-[Fe(NCMe)CO($PPh_2C_6H_4CH{=}NCHPh$-$)_2$][BF_4]$_2$ pre-catalyst [102], were also elaborated, showing, however, lower activity and stereoselectivity.

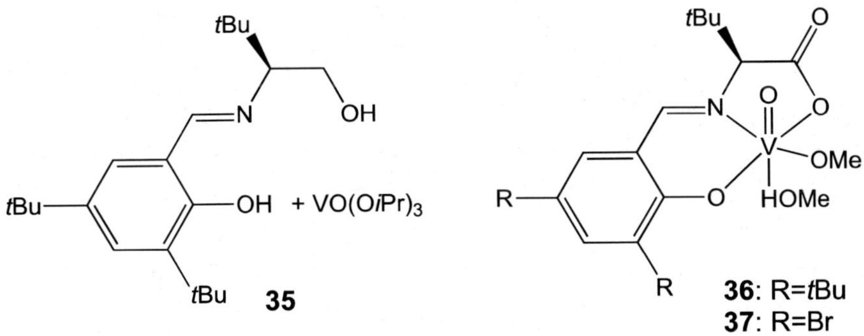

Scheme 11. Vanadium-based catalytic systems.

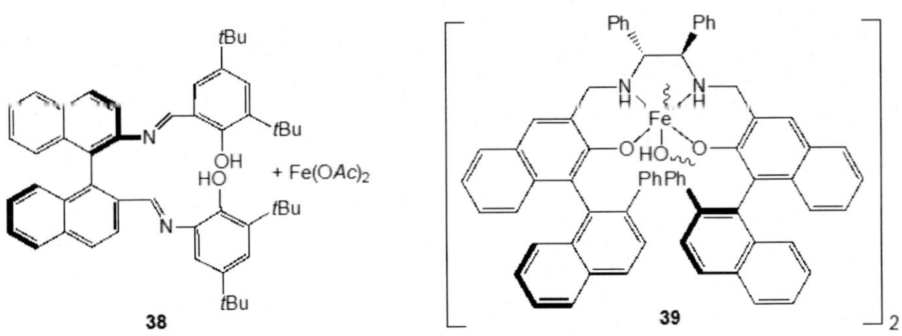

Scheme 12. Iron-based catalysts.

The related to **38** (*R*)-BINAM and quinine-based ligands were also attempted with *zinc* to resolve different types of racemic α-hydroxy ketones [103, 104], but the activities and stereoselectivities achieved were lower then in case of the analogous iron systems.

Along with iron, *copper* is another readily available 3*d* metal, widely used as a catalyst in many oxidation reactions. Accordingly, several works on its application for the OKR of racemic alcohols appeared recently [105–107]. In the first two of them benzoins and amino alcohols were resolved by a $Cu(OTf)_2$–(*R*)-BINAM catalytic system in the presence of TEMPO under molecular oxygen as a terminal oxidant, and enantioselectivities of up to 98% *ee* were reached [105, 106]. Another more recent work applies a chiral nitroxide or hydroxylamine/copper(I) combination for the desymmetrization of axially-prochiral diols (Scheme 13). In such system, the imidazolidinone-based hydroxyl-amine precatalysts are *in situ* transformed into nitroxides which then catalyze the atroposelective aerobic oxidation of the axially-prochiral benzylic diols under mild conditions [107].

Cobalt-catalyzed OKR of *sec*-alcohols has been achieved with molecular oxygen and accompanying olefins (they mediated the oxidation to the corresponding α-ketones). For instance, various secondary benzylic alcohols were subjected to the aerobic oxidative reaction to afford optically active alcohols (enantioselectivities up to 96% *ee*) along with the corresponding ketones by employing chiral ketoiminatocobalt(II) complexes **41**–**43** (Scheme 14) and others such type catalysts [108]. Moreover, the use of a cobalt complex with Schiff base chiral ligands such as **44** (Scheme 14) allowed the OKR of α-hydroxy ketones and α-hydroxy esters by molecular oxygen with good to high (78 to 99.9% *ee*) enantioselectivities [109].

Application of *heterometallic complexes* as catalysts in some cases allows to significantly improve effectivity of the OKR reactions. Thus, $RuCl_2(PPh_3)$(oxazolinylferrocenylphosphine) complexes **45**, **46** (Scheme 15) have been found to be effective catalysts for asymmetric transfer dehydrogenation of racemic *sec*-alcohols with exteremely high enantioselectivity (up to 99.9% *ee*) [42, 110]. Moreover, using a 2:1 mixture of $LaLi_3$-*tris*(binaphthoxide) BINOL/biphenol heterobimetallic complexes

47 and **48** allowed to achieve a kinetic resolution of tertiary nitroaldols with selectivities of 80–97% *ee* and 30–47% recovery yields [111].

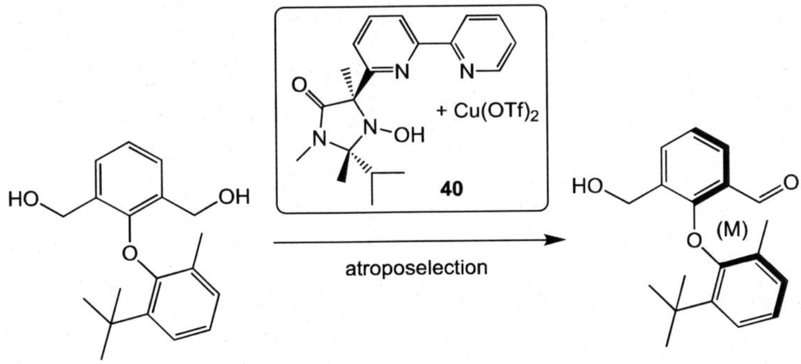

Scheme 13. Atroposelective aerobic oxidation of axially-prochiral benzylic diols.

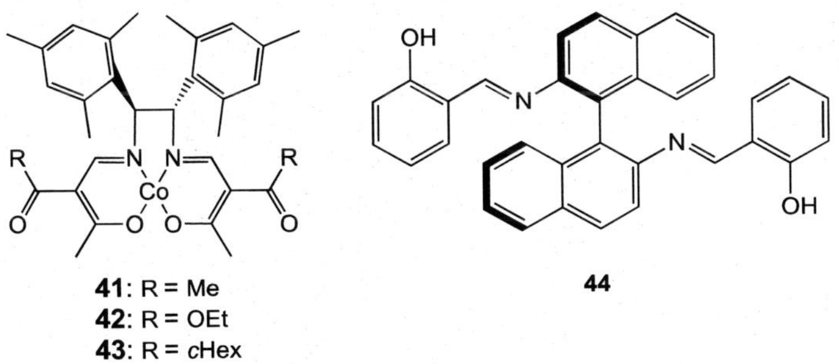

Scheme 14. Some of the catalytic Co-based complexes and ligands.

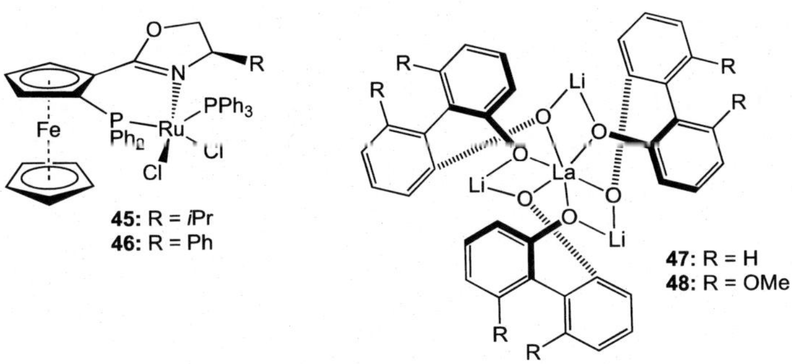

Scheme 15. Ru-Fe and La-Li heterobimetallic complexes suitable for the OKR of alcohols.

Conclusion

Despite the remarkable progress which has been made in the non-enzymatic OKR of racemic alcohols over the past decades, only a few such asymmetric processes have found commercial applications. However, the vector is clearly pointing into this direction, and various secondary alcohols, diols and α-hydroxy acid derivatives have been reliably resolved at the laboratory scale utilizing chiral organocatalysts or transition-metal/ligand systems. It should be noted that most of these systems have limitations in terms of scope, since reliable methods of resolving unactivated alcohols have rarely been reported. Hence, a lot of efforts still should be devoted to this direction, and more general and versatile methods for the OKR of variously functionalized racemic alcohols, such as arylalkyl carbinols, axially-prochiral benzylic diols, cycloalkanols, allylic alcohols, propargylic alcohols, amino alcohols, hydroxy esters, 2-substituted cycloalkanols, 1,2-diol monoesters, 2-amido alcohols, primary as well as tertiary alcohols still have to be developed.

Acknowledgments

This work was supported by the Foundation for Science and Technology (FCT), Portugal (projects IF/01270/2013/CP1163/CT0007, UID/QUI/00100/2013, PTDC/QEQ-QIN/3967/2014, PTDC/QUI-QIN/29778/2017) and "Investigador 2013" (M.N. Kopylovich) Program.

References

[1] Pellissier, Hélène. 2011. "Catalytic non-enzymatic kinetic resolution." *Adv. Synth. Catal.* 353:1613–1666.

[2] Pellissier, Hélène. 2014. "Catalytic kinetic resolution." In *Separation of Enantiomers: Synthetic Methods*, edited by Matthew Todd, 75–122. Weinheim: Wiley-VCH.

[3] Pellissier, Hélène. 2018. "Recent developments in non-enzymatic catalytic oxidative kinetic resolution of secondary alcohols." *Adv. Synth. Catal.* 74: 3459–3468.

[4] Ahmed, Marwa, Tamsin Kelly, and Ashraf Ghanem. 2012. "Applications of enzymatic and non-enzymatic methods to access enantiomerically pure compounds using kinetic resolution and racemization." *Tetrahedron* 68:6781–6802.

[5] Rachwalski, Michał, Niek Vermue, and Floris P. J. T. Rutjes. 2013. "Recent advances in enzymatic and chemical deracemisation of racemic compounds." *Chem. Soc. Rev.*, 42:9268–9282.

[6] Kopylovich, Maximilian N., Ana P. C. Ribeiro, Elisabete C. B. A. Alegria, Nuno M. R. Martins, Luisa M. D. R. S. Martins, and Armando J. L. Pombeiro. 2015. "Catalytic oxidation of alcohols: recent advances." In *Advances in Organometallic Chemistry, ADOMC Volume 63*, edited by Pedro Pérez, 91–174.

[7] Faísca Phillips, Ana M., Armando J. L. Pombeiro, and Maximilian N. Kopylovich. 2017. "Recent advances in cascade reactions initiated by alcohol oxidation." *ChemCatChem* 9:217-246.

[8] Bhat, Vikram, Eric R. Welin, Xuelei Guo, and Brian M. Stoltz. 2017. "Advances in stereoconvergent catalysis from 2005 to 2015: transition-metal-mediated stereoablative reactions, dynamic kinetic resolutions, and dynamic kinetic asymmetric transformations." *Chem. Rev.* 117:4528−4561.

[9] Sigman, Matthew S., and Mitchell J. Schultz. 2004. "The renaissance of palladium(II)-catalyzed oxidation chemistry." *Org. Biomol. Chem.*, 2:2551−2554.

[10] Sigman, Matthew S., and David R. Jensen. 2006. "Ligand-modulated palladium-catalyzed aerobic alcohol oxidations." *Acc. Chem. Res.* 39:221−229.

[11] Bartoszewicz, Agnieszka, Nanna Ahlsten, and Belén Martín-Matute. 2013. "Enantioselective synthesis of alcohols and amines by iridium-

catalyzed hydrogenation, transfer hydrogenation, and related processes." *Chem. Eur. J.* 19:7274–7302.

[12] Bryliakov, Konstantin. 2015. *Environmentally Sustainable Catalytic Asymmetric Oxidations*. CRC Press, Taylor & Francis Group.

[13] Irie, Ryo, Tatsuya Uchida, and Kazuhiro Matsumoto. 2015. "Katsuki catalysts for asymmetric oxidation: design concepts, serendipities for breakthroughs, and applications." *Chem. Lett.*, 44:1268–1283.

[14] Ahmad, Irshad, Shagufta, and Abdul R. AlMallah. 2017. "Advances in asymmetric oxidative kinetic resolution of racemic secondary alcohols catalyzed by chiral Mn(III) salen complexes." *Chirality*. 29:798–810.

[15] Yang, Hui, and Wen-Hua Zheng. 2018. "Recent advances on nonenzymatic catalytic kinetic resolution of diols." *Tetrahedron Lett.* 59:583-591.

[16] Berti, Corrado, and John M. Perkins. 1979. "Enantioselective oxidation by a chiral acylaminyl oxide." *Angew. Chem., Int. Ed.* 18:864–865.

[17] Rychnovsky, Scott D.; McLernon, Terri L.; Rajapakse, Hemaka. 1996. "Enantioselective oxidation of secondary alcohols using a chiral nitroxyl (N-oxoammonium salt) catalyst." *J. Org. Chem.* 61:1194–1195.

[18] Kashiwagi, Yoshitomo, Futoshi Kurashima, Chikara Kikuchi, Jun-ichi Anzai, Tetsuo Osa, and James M. Bobbitt. 1999. "Enantioselective electrocatalytic oxidation of racemic alcohols using a chiral 1-azaspiro[5.5]undecane-N-oxyl radical." *Tetrahedron Lett.* 40:6469–6472.

[19] Kashiwagi, Yoshitomo, Yoshinori Yanagisawa, Futoshi Kurashima, Jun-ichi Anzai, Tetsuo Osa, and James M. Bobbitt. 1996. "Enantioselective electrocatalytic oxidation of racemic alcohols on a TEMPO-modified graphite felt electrode by use of chiral base (TEMPO = 2,2,6,6-tetramethylpiperidin-1-yloxyl)." *Chem. Commun.* 2745–2746.

[20] Shiigi, Hirofumi, Hiroyuki Mori, Tomoaki Tanaka, Yosuke Demizu, and Osamu Onomura. 2008. "Chiral azabicyclo-N-oxyls mediated

enantioselective electrooxidation of sec-alcohols." *Tetrahedron Lett.* 49:5247–5251.

[21] Tomizawa, Masaki, Masatoshi Shibuya, and Yoshiharu Iwabuchi. 2009. "Highly enantioselective organocatalytic oxidative kinetic resolution of secondary alcohols using chirally modified AZADOs." *Org. Lett.* 11:1829–1831.

[22] Tomizawa, Masaki, Masatoshi Shibuya, and Yoshiharu Iwabuchi. 2014. "Correction to Highly enantioselective organocatalytic oxidative kinetic resolution of secondary alcohols using chirally modified AZADOs." *Org. Lett.* 16:4968–4968.

[23] Hamada, Shohei, Yoshiyuki Wada, Takahiro Sasamori, Norihiro Tokitoh, Takumi Furura, and Takeo Kawabata. 2014. "Oxidative kinetic resolution of racemic alkyl aryl carbinols." *Tetrahedron Lett.* 55:1943−1945.

[24] Formaggio, Fernando, Marcella Bonchio, Marco Crisma, Cristina Peggion, Stefano Mezzato, Alessandra Polese, Alessandra Barazza, Sabrina Antonello, Flavio Maran, Quirinus B. Broxterman, Bernard Kaptein, Johan Kamphuis, Rosa Maria Vitale, Michele Saviano, Ettore Benedetti, and Claudio Toniolo. 2002. "Nitroxyl peptides as catalysts of enantioselective oxidations." *Chem. Eur. J.* 8:84−93.

[25] Graetz, Benjamin, Scott Rychnovsky, Wen-Hao Leu, Patrick Farmer, and Rong Lin. 2005. "C2-Symmetric nitroxides and their potential as enantioselective oxidants" *Tetrahedron Asymm.* 16:3584−3598.

[26] Maayan, Galia, Michael D. Ward, and Kent Kirshenbaum. 2009. "Folded biomimetic oligomers for enantioselective catalysis." *Proc. Natl. Acad. Sci. U. S. A.* 106:13679−13684.

[27] Kashiwagi, Yoshitomo, Futoshi Kurashima, Shinya Chiba, Jun-ichi Anzai, Tetsuo Osa, and James M. Bobbitt. 2003. "Asymmetric electrochemical lactonization of diols on a chiral 1-azaspiro[5.5]undecane N-oxyl radical mediator-modified graphite felt electrode." *Chem. Commun.* 114−115.

[28] Carbó López, Marta, Guy Royal, Christian Philouze, Pierre Y. Chavant, and Véronique Blandin. 2014. "Imidazolidinone nitroxides as catalysts in the aerobic oxidation of alcohols, en route to

atroposelective oxidative desymmetrization." *Eur. J. Org. Chem.* 4884–4896.

[29] Murakami, Keiichi, Yusuke Sasano, Masaki Tomizawa, Masatoshi Shibuya, Eunsang Kwon, and Yoshiharu Iwabuchi. 2014. "Highly enantioselective organocatalytic oxidative kinetic resolution of secondary alcohols using chiral alkoxyamines as precatalysts: catalyst structure, active species, and substrate scope." *J. Am. Chem. Soc.* 136:17591–17600.

[30] Yamakoshi, Hiroyuki, Masatoshi Shibuya, Masaki Tomizawa, Yuji Osada, Naoki Kanoh, and Yoshiharu Iwabuchi. 2010. "Total synthesis and determination of the absolute configuration of (−)-idesolide." *Ogr. Lett.* 12:980–983.

[31] Sekar, Govindasamy, and Hisao Nishiyama. 2001. "Nonenzymatic kinetic resolution of secondary alcohols: enantioselective SN2 displacement of hydroxy groups by halogens in the presence of chiral BINAP." *J. Am. Chem. Soc.* 123:3603–3604.

[32] Rong, Zi-Qiang, Hui-Jie Pan, Hai-Long Yan, and Yu Zhao. 2014. "Enantioselective oxidation of 1,2-diols with quinine-derived urea organocatalyst." *Org. Lett.* 16:208–211.

[33] Luo, Sanzhong, Pengxin Zhou, Jiuyuan Li, and Jin-Pei Cheng. 2010. "Asymmetric retro- and transfer-aldol reactions catalyzed by a simple chiral primary amine." *Chem. Eur. J.* 16:4457–4461.

[34] D'Accolti, Lucia, Antonia Detomaso, Caterina Fusco, Angela Rosa, and Ruggero Curci. 1993. "Selective oxidation of optically active *sec,sec*-1,2-diols by dioxiranes. A practical method for the synthesis of homochiral α-hydroxy ketones in high optical purity." *J. Org. Chem.* 58:3600–3601.

[35] Adam, Waldemar, Chantu R. Saha-Möller, and Cong-Gui Zhao. 1998. "Enantioselective oxidation of *vic*-diols to optically active α-hydroxy ketones by a fructose-derived dioxirane." *Tetrahedron Asymm.* 9:4117–4122.

[36] Adam, Waldemar, Chantu R. Saha-Möller, and Cong-Gui Zhao. 1999. "Asymmetric C-H oxidation of *vic*-diols to α-hydroxy ketones by a

fructose-derived dioxirane: electronic effects on the enantioselectivity of oxygen transfer." *J. Org. Chem.* 64:7492−7497.

[37] Jakka, Kavitha, and Cong-Gui Zhao. 2006. "Highly enantioselective C−H oxidation of *vic*-diols with Shi's oxazolidinone dioxiranes." *Org. Lett.* 8:3013−3015.

[38] Allen, Anna E., and David W. C. MacMillan. 2012. "Synergistic catalysis: a powerful synthetic strategy for new reaction development." *Chem. Sci.* 3:633−658.

[39] Müller, Christian E., Radim Hrdina, Raffael C. Wende, and Peter R. Schreiner. 2011. "A multicatalyst system for the one-pot desymmetrization/oxidation of meso-1,2-alkane diols." *Chem. Eur. J.* 17:6309−6314.

[40] Parmeggiani, Camilla, and Francesca Cardona. 2012. "Transition metal based catalysts in the aerobic oxidation of alcohols." *Green Chem.* 14:547−564.

[41] Hashiguchi, Shohei, Akio Fujii, Karl-Josef Haack, Kazuhiko Matsumura, Takao Ikariya, and Ryoji Noyori. 1997. "Kinetic resolution of racemic secondary alcohols by Ru^{II}-catalyzed hydrogen transfer." *Angew. Chem. Int. Ed.* 36:288–290.

[42] Nishibayashi, Yoshiaki, Akiyoshi Yamauchi, Gen Onodera, and Sakae Uemura. 2003. "Oxidative kinetic resolution of racemic alcohols catalyzed by chiral ferrocenyloxazolinylphosphine-ruthenium complexes." *J. Org. Chem.* 68:5875–5880.

[43] Tan, Duan-Ming, and Kin Shing Chan. 2005. "Asymmetric transfer hydrogenation of ketones in 2-propanol catalyzed by arsinooxazoline–ruthenium(II) complex." *Tetrahedron Lett.* 46:503–505.

[44] Masutani, Kouta, Tatsuya Uchida, Ryo Irie, and Tsutomu Katsuki. 2000. "Catalytic asymmetric and chemoselective aerobic oxidation: kinetic resolution of *sec*-alcohols." *Tetrahedron Lett.* 41:5119−5123.

[45] Shimizu, Hideki, Kenya Nakata, and Tsutomu Katsuki. 2002. "(Salen)ruthenium-catalyzed desymmetrization of meso-diols: catalytic aerobic asymmetric oxidation under photoirradiation." *Chem. Lett.* 31:1080–1081.

[46] Shimizu, Hideki, and Tsutomu Katsuki. 2003. "(Salen)ruthenium-catalyzed desymmetrization of meso-diols (2): apical ligand effect on enantioselectivity." *Chem. Lett.* 32:480–481.

[47] Shimizu, Hideki, Satoaki Onitsuka, Hiromichi Egami, and Tsutomu Katsuki. 2005. "Ruthenium(salen)-catalyzed aerobic oxidative desymmetrization of meso-diols and its kinetics." *J. Am. Chem. Soc.* 127:5396–5413.

[48] Nakamura, Yukie, Hiromichi Egami, Kazuhiro Matsumoto, Tatsuya Uchida, and Tsutomu Katsuki. 2007. "Aerobic oxidative kinetic resolution of racemic alcohols with bidentate ligand-binding Ru(salen) complex as catalyst." *Tetrahedron*. 63:6383–6387.

[49] Mizoguchi, Hirotaka, Tatsuya Uchida, Kohichi Ishida, and Tsutomu Katsuki. 2009. "Ru(PPh_3)(OH)-salen complex: a designer catalyst for chemoselective aerobic oxidation of primary alcohols." *Tetrahedron Lett*. 50:3432–3435.

[50] Mizoguchi, Hirotaka, Tatsuya Uchida, and Tsutomu Katsuki. 2014. "Ruthenium-catalyzed oxidative kinetic resolution of unactivated and activated secondary alcohols with air as the hydrogen acceptor at room temperature." *Angew. Chem. Int. Ed.* 53:3178–3182.

[51] Jensen, David R., Jacob S. Pugsley, and Matthew S. Sigman. 2001. "Palladium-catalyzed enantioselective oxidations of alcohols using molecular oxygen." *J. Am. Chem. Soc.* 123:7475–7476.

[52] Eric M. Ferreira, and Brian M. Stoltz. 2001. "The palladium-catalyzed oxidative kinetic resolution of secondary alcohols with molecular oxygen." *J. Am. Chem. Soc.* 123:7725–7726.

[53] Mueller, Jaime A., David R. Jensen, and Matthew S. Sigman. 2002. "Dual role of (−)-sparteine in the palladium-catalyzed aerobic oxidative kinetic resolution of secondary alcohols." *J. Am. Chem. Soc.* 124:8202–8203.

[54] Bagdanoff, Jeffrey T., Eric M. Ferreira, and Brian M. Stoltz. 2003. "Palladium-catalyzed enantioselective oxidation of alcohols: a dramatic rate acceleration by Cs_2CO_3/*t*-BuOH." *Org. Lett.* 5:835–837.

[55] Mandal, Sunil K., David R. Jensen, Jacob S. Pugsley, and Matthew S. Sigman. 2003. "Scope of enantioselective palladium(II)-catalyzed

aerobic alcohol oxidations with (−)-sparteine." *J. Org. Chem.* 68:4600–4603.

[56] Mandal, Sunil K., and Matthew S. Sigman. 2003. "Palladium-catalyzed aerobic oxidative kinetic resolution of alcohols with an achiral exogenous base." *J. Org. Chem.* 68: 7535–7537.

[57] Mueller Jaime A., and Matthew S. Sigman. 2003. "Mechanistic investigations of the palladium-catalyzed aerobic oxidative kinetic resolution of secondary alcohols using (-)-sparteine." *J. Am. Chem. Soc.* 125: 7005–7013.

[58] Jensen, David R., and Matthew S. Sigman. 2003. "Palladium catalysts for aerobic oxidative kinetic resolution of secondary alcohols based on mechanistic insight." *Org. Lett.* 5:63–65.

[59] Trend, Raissa M., and Brian M. Stoltz. 2004. "An experimentally derived model for stereoselectivity in the aerobic oxidative kinetic resolution of secondary alcohols by (sparteine)$PdCl_2$." *J. Am. Chem. Soc.* 126: 4482–4483.

[60] Ebner, David C., Jeffrey T. Bagdanoff, Eric M. Ferreira, Ryan M. McFadden, Daniel D. Caspi, Raissa M. Trend, and Brian M. Stoltz. 2009. "The palladium-catalyzed aerobic kinetic resolution of secondary alcohols: reaction development, scope, and applications." *Chem. Eur. J.* 15:12978–12992.

[61] Meyer, Michael E., John H. Phillips, Eric M. Ferreira, and Brian M. Stoltz. 2013. "Use of a palladium(II)-catalyzed oxidative kinetic resolution in synthetic efforts toward bielschowskysin." *Tetrahedron.* 69:7627–7635.

[62] Ebner, David C., Raissa M. Trend, Cédric Genet, Matthew J. McGrath, Peter O'Brien, and Brian M. Stoltz. 2008. "Palladium-catalyzed enantioselective oxidation of chiral secondary alcohols: access to both enantiomeric series." *Angew. Chem. Int. Ed.* 47:6367–6370.

[63] Caspi, Daniel D., David C. Ebner, Jeffrey T. Bagdanoff, and Brian M. Stoltz. 2008. "Resolution of important pharmaceutical building blocks by palladium-catalyzed aerobic oxidation of secondary alcohols." *Adv. Synth. Catal.* 346:185–189.

[64] Tambar, Uttam K., David C. Ebner, and Brian M. Stoltz. 2006. "A convergent and enantioselective synthesis of (+)-amurensinine via selective C–H and C–C bond insertion reactions." *J. Am. Chem. Soc.* 128:11752–11753.

[65] Trend, Raissa M., and Brian M. Stoltz. 2008. "Structural features and reactivity of (sparteine)$PdCl_2$: a model for selectivity in the oxidative kinetic resolution of secondary alcohols." *J. Am. Chem. Soc.* 130:15957–15966.

[66] Dearden, Michael J., Matthew J. McGrath, and Peter O'Brien. 2004. "Evaluation of (+)-sparteine-like diamines for asymmetric synthesis." *J. Org. Chem.* 69:5789–5792.

[67] Breuning, Matthias, Melanie Steiner, Christian Mehler, Alexander Paasche, and David Hein. 2009. "A flexible route to chiral 2-*endo*-substituted 9-oxabispidines and their application in the enantioselective oxidation of secondary alcohols." *J. Org. Chem.* 74:1407–1410.

[68] Lesma, Giordano, Tullio Pilati, Alessandro Sacchetti, and Alessandra Silvani. 2008. "New chiral diamino ligands as sparteine analogues: application to the palladium-catalyzed kinetic oxidative resolution of 1-phenyl ethanol." *Tetrahedron Asymm.* 19:1363–1366.

[69] Chen, Tao, Jia-Jun Jiang, Qin Xu, and Min Shi. 2007. "Axially chiral NHC–Pd(II) complexes in the oxidative kinetic resolution of secondary alcohols using molecular oxygen as a terminal oxidant." *Org. Lett.* 9:865–868.

[70] Liu, Shi-Jia, Lian-jun Liu, and Min Shi. 2009. "Preparation of novel axially chiral NHC-Pd(II) complexes and their application in oxidative kinetic resolution of secondary alcohols." *Appl. Organomet. Chem.* 23:183–190.

[71] Sun, Wei, Hongwang Wang, Chungu Xia, Jingwei Li, and Peiqing Zhao. 2003. "Chiral-Mn(salen)-complex-catalyzed kinetic resolution of secondary alcohols in water." *Angew. Chem. Int. Ed.* 42:1042–1044.

[72] Li, Zhen, Zhong H. Tang, Xiao X. Hu, and Chun G. Xia. 2005. "Insight into the mechanism of oxidative kinetic resolution of racemic

secondary alcohols by using manganese(III)(salen) complexes as catalysts." *Chem. Eur. J.* 11:1210–1216.

[73] Han, Furong, Jiquan Zhao, Yuecheng Zhang, Weiyu Wang, Yanyan Zuo, and Junwei An. 2008. "Oxidative kinetic resolution of 1-phenylethanol catalyzed by sugar-based salen–Mn(III) complexes." *Carbohydrate Res.* 343:1407–1413.

[74] Cheng, Qigan, Fanguo Deng, Chungu Xia, and Wei Sun. 2008. "The Mn(Salen)-catalyzed oxidative kinetic resolution of secondary alcohols: reaction development and scope." *Tetrahedron Asymm.* 19:2359–2362.

[75] Kantam, Lakshmi M., T. Ramani, L. Chakrapani, and B. M. Choudary. 2007. "Oxidative kinetic resolution of racemic secondary alcohols catalyzed by resin supported sulfonato-Mn(salen) complex in water." *J. Mol. Cat. A Chem.* 274:11–15.

[76] Pathak, Kavita, Irshad Ahmad, Sayed H. R. Abdi, Rukhsana I. Kureshy, Noorul H. Khan, and Raksh V. Jasra. 2007. "Oxidative kinetic resolution of racemic secondary alcohols catalyzed by recyclable chiral dimeric Mn(III) salen catalysts." *J. Mol. Cat. A Chem.* 274:120–126.

[77] Sahoo, Suman, Pradeep Kumar, F. Lefebvre, and S. B. Halligudi. 2009. "Oxidative kinetic resolution of alcohols using chiral Mn–salen complex immobilized onto ionic liquid modified silica." *Appl. Cat. A Gen.* 354:17–25.

[78] Li, Chengyong, Jiangfeng Zhao, Rong Tan, Zhigang Peng, Rongchang Luo, Ming Peng, and Donghong Yin. 2011. "Recyclable ionic liquid-bridged chiral dimeric salen Mn(III) complexes for oxidative kinetic resolution of racemic secondary alcohols." *Catal. Commun.* 15:27–31.

[79] Bera, Prasanta Kumar, Nabin Ch. Maity, Sayed H. R. Abdi, Noor-ul H. Khan, Rukhsana I. Kureshy, and Hari C. Bajaj. 2013. "Macrocyclic Mn(III) salen complexes as recyclable catalyst for oxidative kinetic resolution of secondary alcohols." *Appl. Cat. A Gen.* 467: 542–551.

[80] Maity, Nabin Ch. Prasanta Kumar Bera, S. Saravanan, Sayed H. R. Abdi, Rukhsana I. Kureshy, Noor-ul H. Khan, and Hari C. Bajaj.

2014. "Diethyl tartrate linked chiral macrocyclic manganese(III)–salen complex for enantioselective epoxidation of olefins and oxidative kinetic resolution of racemic secondary alcohols." *Chem Plus Chem*, 79:1426–1433.

[81] Xu, Daqian, Shoufeng Wang, Zhiqiang Shen, Chungu Xia, and Wei Sun. 2012. "Enantioselective oxidation of racemic secondary alcohols catalyzed by chiral Mn(III)-salen complexes with N-bromosuccinimide as a powerful oxidant." *Org. Biomol. Chem.* 10:2730–2732.

[82] Miao, Chengxia, Xiao-Xi Li, Yong-Min Lee, Chungu Xia, Yong Wang, Wonwoo Nam, and Wei Sun. 2014. "Enantioselective oxidation of racemic secondary alcohols catalyzed by chiral Mn(III)-salen complex with sodium hypochlorite as oxidant." *Catal. Commun.* 45:114–117.

[83] Miao, Chengxia, Xiao-Xi Li, Yong-Min Lee, Chungu Xia, Yong Wang, Wonwoo Nam, and Wei Sun. 2017. "Manganese complex-catalyzed oxidation and oxidative kinetic resolution of secondary alcohols by hydrogen peroxide." *Chem. Sci.* 8:7476–7482.

[84] Brown, M. Kevin, Megan M. Blewett, James R. Colombe, and E. J. Corey. 2010. "Mechanism of the enantioselective oxidation of racemic secondary alcohols catalyzed by chiral Mn(III)-salen complexes." *J. Am. Chem. Soc.* 132:1116–1117.

[85] Talsi, Evgenii P., Denis G. Samsonenko, and Konstantin P. Bryliakov. 2017. "Asymmetric autoamplification in the oxidative kinetic resolution of secondary benzylic alcohols catalyzed by manganese complexes." *ChemCatChem*. 9:2599–2607.

[86] Talsi, Evgenii P., and Konstantin P. Bryliakov. 2018. "Autoamplification-enhanced oxidative kinetic resolution of sec-alcohols and alkyl mandelates, and its kinetic model." *Chem Cat Chem*. 10: 2693–2699.

[87] Li, Yan-Yun, Xue-Qin Zhang, Zhen-Rong Dong, Wei-Yi Shen, Gui Chen, and Jing-Xing Gao. 2006. "Kinetic resolution of racemic secondary alcohols catalyzed by chiral diaminodiphosphine–Ir(I) complexes." *Org. Lett.* 8:5565–5567.

[88] Zhang, Juanni, Xiangren Yang, Han Zhou, Yanyun Li, Zhenrong Dong, and Jingxing Gao. 2012. "Oxidative kinetic resolution of racemic secondary alcohols in water with chiral PNNP/Ir catalyst." *Green Chem.* 14:1289–1292.

[89] Arita, Sachiko, Takashi Koike, Yoshihito Kayaki, and Takao Ikariya. 2008. "Aerobic oxidative kinetic resolution of racemic secondary alcohols with chiral bifunctional amido complexes." *Angew. Chem. Int. Ed.* 47:2447–2449.

[90] Suzuki, Takeyuki, Kazem Ghozati, Tadashi Katoh, and Hiroaki Sasai. 2009. "Ir-catalyzed oxidative desymmetrization of *meso*-diols." *Org. Lett.*, 11:4286–4288.

[91] Moritani, Junki, Yasuharu Hasegawa, Yoshihito Kayaki, and Takao Ikariya. 2014. "Aerobic oxidative desymmetrization of *meso*-diols with bifunctional amidoiridium catalysts bearing chiral N-sulfonyldiamine ligands." *Tetrahedron Lett.* 55:1188–1191.

[92] Radosevich, Alexander T., Christine Musich, and F. Dean Toste. 2005. "Vanadium-catalyzed asymmetric oxidation of α-hydroxy esters using molecular oxygen as stoichiometric oxidant." *J. Am. Chem. Soc.* 127:1090–1091.

[93] Pawar, Vijay D., Sampada Bettigeri, Shiue-Shien Weng, Jun-Qi Kao, and Chien-Tien Chen. 2006. "Highly enantioselective aerobic oxidation of α-hydroxyphosphonates catalyzed by chiral vanadyl(V) methoxides bearing N-salicylidene-α-aminocarboxylates." *J. Am. Chem. Soc.* 128:6308–6309.

[94] Chen, Chien-Tien, Sampada Bettigeri, Shiue-Shien Weng, Vijay D. Pawar, Ya-Hui Lin, Cheng-Yuan Liu, and Way-Zen Lee. 2007. "Asymmetric aerobic oxidation of α-hydroxy acid derivatives by C_4-symmetric, vanadate-centered, tetrakisvanadyl(V) clusters derived from N-salicylidene-α-aminocarboxylates." *J. Org. Chem.*, 72:8175–8185.

[95] Chen, Chien-Tien, Jun-Qi Kao, Santosh B. Salunke, and Ya-Hui Lin. 2011. "Enantioselective aerobic oxidation of α-hydroxy-ketones catalyzed by oxidovanadium(V) methoxides bearing chiral, N-salicylidene-tert-butylglycinates." *Org. Lett.* 13:26–29.

[96] Weng, Shiue-Shien, Mei-Wen Shen, Jun-Qi Kao, Yogesh S. Munot, and Chien-Tien Chen. 2006. "Chiral N-salicylidene vanadyl carboxylate-catalyzed enantioselective aerobic oxidation of α-hydroxy esters and amides." *Proc. Natl. Acad. Sci. U.S.A.* 103:3522–3527.

[97] Radosevich, Alexander T., Vincent S. Chan, Hui-Wen Shih, and F. Dean Toste. 2008. "Synthesis of (–)-octalactin a by a strategic vanadium-catalyzed oxidative kinetic resolution." *Angew. Chem. Int. Ed.*, 47:3755–3758.

[98] Yin, Lu, Xian Jia, Xing Shu Li, Albert S. C. Chan. 2010. "Simply air: vanadium-catalyzed oxidative kinetic resolution of methyl o-chloromandelate by ambient air." *Chin. Chem. Lett.* 21:774–777.

[98] Yin, Lu, Xian Jia, Xing Shu Li, Albert S. C. Chan. 2010. "Simply air: vanadium-catalyzed oxidative kinetic resolution of methyl o-chloromandelate by ambient air." *Chin. Chem. Lett.* 21:774–777.

[99] Shiels, Rebecca A., Krishnan Venkatasubbaiah, and Christopher W. Jones. 2008. "Polymer- and silica-supported tridentate Schiff base–vanadium catalysts for asymmetric oxidation of ethyl mandelate: activity, stability and recyclability." *Adv. Synth. Catal.* 350:2823–2834.

[100] Muthupandi, Pandi, Santosh Kumar Alamsetti, and Govindasamy Sekar. 2008. "Chiral iron complex catalyzed enantioselective oxidation of racemic benzoins." *Chem. Commun.* 3288–3290.

[101] Kunisu, Takashi, Takuya Oguma, and Tsutomu Katsuki. 2011. "Aerobic oxidative kinetic resolution of secondary alcohols with naphthoxide-bound iron(salan) complex." *J. Am. Chem. Soc.* 133:12937–12939.

[102] Sonnenberg, Jessica F., Dmitry Pichugin, Neil Coombs, and Robert H. Morris. 2013. "Oxidative kinetic resolution of aromatic alcohols using iron nanoparticles." *Top. Catal.* 56:1199–1207.

[103] Muthupandi, Pandi, and Govindasamy Sekar. 2011. "Zinc-catalyzed aerobic oxidation of benzoins and its extension to enantioselective oxidation." *Tetrahedron Lett.* 52:692–695.

[104] Muthupandi, Pandi, and Govindasamy Sekar. 2011. "Chiral Zn-catalyzed aerobic oxidative kinetic resolution of α-hydroxy ketones." *Tetrahedron Asymm.* 22:512–517.

[105] Alamsetti, Santosh Kumar, Sreedevi Mannam, Pandi Mutupandi, and Govindasamy Sekar. 2009. "Galactose oxidase model: biomimetic enantiomer-differentiating oxidation of alcohols by a chiral copper complex." *Chem. Eur. J.* 15:1086–1090.

[106] Mannam, Sreedevi, and Govindasamy Sekar. 2009. "An enantiopure galactose oxidase model: synthesis of chiral amino alcohols through oxidative kinetic resolution catalyzed by a chiral copper complex." *Tetrahedron Asymm.* 20:497–502.

[107] López, Marta Carbó, Pierre Y. Chavant, Florian Molton, Guy Royal, and Véronique Blandin. 2017. "Chiral nitroxide/copper-catalyzed aerobic oxidation of alcohols: atroposelective oxidative desymmetrization." *Chemistry Select.* 2:443–450.

[108] Yamada, Tohru, Sho Higano, Takanori Yano, and Yoshihiro Yamashita. 2009. "Cobalt-catalyzed oxidative kinetic resolution of secondary benzylic alcohols with molecular oxygen." *Chem. Lett.* 38:40–41.

[109] Alamsetti, Santosh Kumar, and Govindasamy Sekar. 2010. "Chiral cobalt-catalyzed enantioselective aerobic oxidation of α-hydroxy esters." *Chem. Commun.* 7235–7237.

[110] Nishibayashi, Yoshiaki, Izuru, Takei, Sakae Uemura, and Masanobu Hidai. 1999. "Extremely high enantioselective redox reaction of ketones and alcohols catalyzed by $RuCl_2(PPh_3)$(oxazolinylferrocenyl phosphine)." *Organometallics.* 18:2291–2293.

[111] Tosaki, Shin-ya, Keiichi Hara, Vijay Gnanadesikan, Hiroyuki Morimoto, Shinji Harada, Mari Sugita, Noriyuki Yamagiwa, Shigeki Matsunaga, and Masakatsu Shibasaki. 2006. "Mixed La-Li heterobimetallic complexes for tertiary nitroaldol resolution." *J. Am. Chem. Soc.* 128: 11776–11777.

In: Alcohol Oxidation
Editor: Steffan Paterson

ISBN: 978-1-53614-604-2

Chapter 5

ALCOHOL ELECTROOXIDATION: A RENEWABLE SOURCE OF ENERGY AND CHEMICALS

Antonio C. D. Ângelo and Francielle Bortoloti*
Electrocatalysis Laboratory, UNESP, Bauru, SP, Brazil

ABSTRACT

In this chapter, the potential employment of the electrochemical process to oxidize alcohols in a more efficient way than the usual chemical method is presented. The use of the electrode potential as a controllable variable and driving force of the process allows for the oxidation of some alcohol molecules without of the very aggressive nature of reactants. The use of alcohols as fuel in fuel cells opens a wide field of improvements for energy and added value chemicals production. Methanol and ethanol electrooxidation are discussed and very promising results obtained from ethylene glycol and glycerol reactions show the possible use of these alcohols as precursors for energy production in a sustainable development concept as well as the production of chemicals of interest for industries.

* Corresponding Author Email: acangelo@fc.unesp.br.

1. CHEMICAL ALCOHOL OXIDATION

Alcohols are organic compounds with various applications, including energy production and synthesis of countless chemical products.

Oxidation is certainly the best-known chemical reaction. In contact with several sorts of oxidizing agents, alcohols may generate carbonyl compounds (aldehyde, ketone, and carboxylic acids) depending on the oxidizing power of said agents. Different chemical reactants can act as oxidizers; e.g., CrO_3, $Na_2Cr_2O_7$, and $KMnO_4$, in acidic (H_2SO_4) solution. The choice of oxidizing agent will depend on the desired product, costs, practical convenience, and reaction yield, among other factors. The pristine alcohol structure will influence the main product of each reaction: a primary alcohol will afford aldehyde or carboxylic acid, a secondary alcohol will give ketone, and a tertiary alcohol will not react with most oxidizing agents. However, the use of chromium derivatives as oxidizers may cause considerable environmental damage, which has called for alternatives to the use of such polluting agents. In this sense, the Swern oxidation path comes into play [1]. This oxidation path consists of using dimethylsulfoxide [$(CH_3)_2SO$], oxyalyl chloride ($COCl_2$), and trimethylamine ($(CH_3CH_2)_3N$) compounds; the dimethyl-chlor-sulfonic ion ($(CH_3)_2SCl)^+$) underlies the oxidation process.

In the next items, we will examine these reactions in more detail.

1.1. Primary Alcohol

Primary alcohol is initially oxidized to aldehyde. Nevertheless, the reaction continues, and the aldehyde is oxidized to carboxylic acid, as depicted in reaction [1].

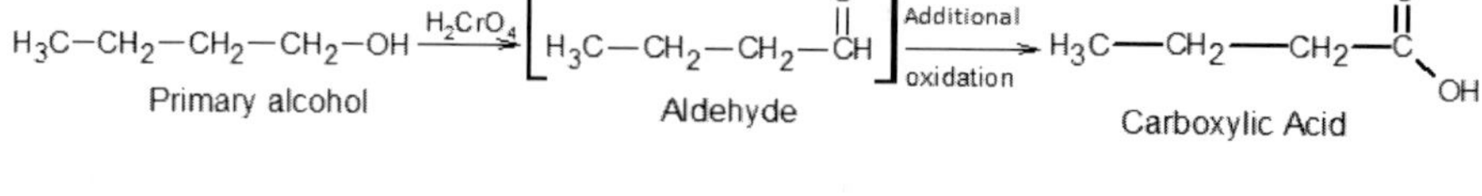

The use of pyridine chloro-chromate (PCC) as oxidizing agent together with an anhydrous solvent such as dichloromethane can avoid additional oxidation.

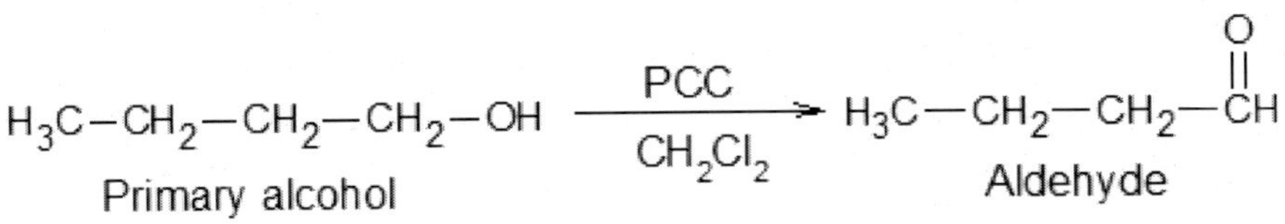

Because the reaction does not take place in non-aqueous medium, primary alcohol oxidation through the Swern path can also stop at the aldehyde.

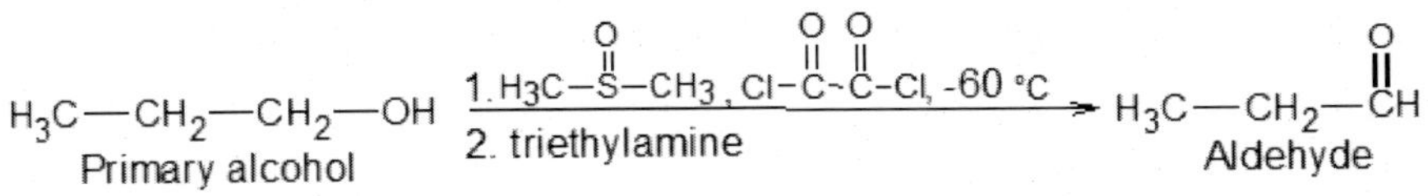

1.2. Secondary Alcohol

Secondary alcohols are quickly oxidized to ketones in great yield [1]:

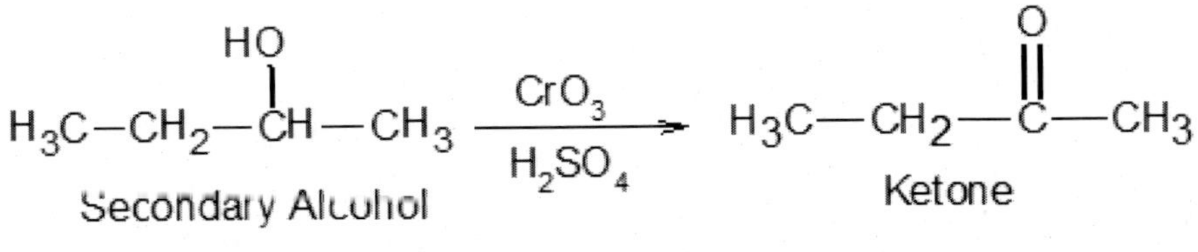

Secondary alcohols can also be oxidized through the Swern path:

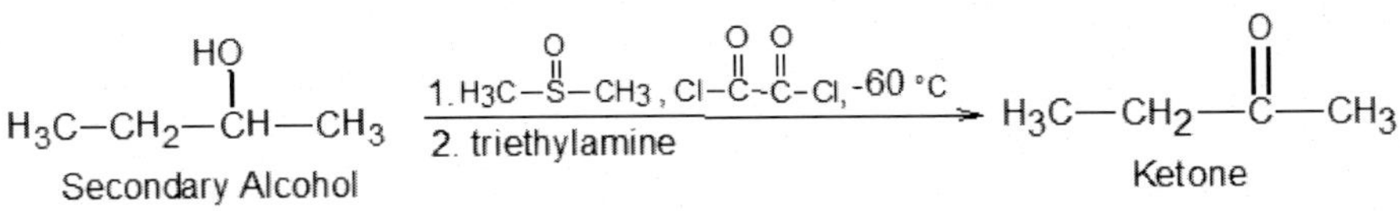

1.3. Tertiary Alcohol

Given that alcohol oxidation involves hydrogen removal from the carbon to which OH is bound, it is easy to understand why tertiary alcohols are not oxidized to a carbonyl group [1]:

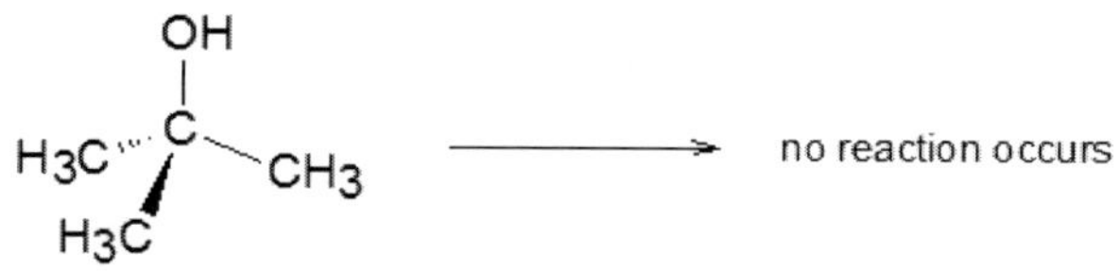

1.4. The Green Chemistry Alternative for Alcohol Oxidation

The usual chemical alcohol oxidations cited earlier generate a considerable amount of dangerous residues, which are discharged as heavy metal compounds into the environment. Technological advances and global concern about the impact of polluting agents on the environment have led innovative and cleaner chemical alcohol oxidation methodologies to be investigated and developed. Among these methodologies, we can cite alcohol oxidation in the presence of molecular oxygen, TEMPO (2,2,6,6-tetramethylpiperidinyl-1-oxy), and hydrogen peroxide.

Alcohol oxidation by molecular oxygen can be conducted in homogeneous systems; Pd, Cu, or Ru compounds can be used, and toluene can be employed as solvent. Brink and co-workers [2] stated that if such reaction were performed in aqueous medium, it could be considerably safer and less costly, not to mention that water is an environmentally friendly medium, which is one of the aims of green chemistry.

Li et al. [3] used different organic solvents (toluene; α,α,α -trifluorotoluene; bromobenzene; benzene; and chlorobenzene) and heterogeneous catalysts like $MnFe_{1.5}Ru_{0.35}Cu_{0.15}O_4$ and showed that these compounds efficiently converted alcohols (alkenic, benzylic, secondary, primary) to carbonyl compounds (aldehydes and ketones). According to these authors, coordinated unsaturation of the isolated species R=O underlies this oxidation reaction. Moreover, these authors enumerated numerous advantages of the employed method: air is the only oxidizing agent; the reactions occur at room temperature; the catalyst can be recycled; and distinct alcohols can be applied.

Other studies developed with noble metals including Pt and Pd supported on carbon are well known in the field of catalyzed aerobic alcohol oxidation reactions in aqueous medium. However, solubility in water limits the use of this method. Brink [2] stated that aerobic oxidation by activated and non-activated hydroxyl groups is highly efficient whenever the soluble o-phenanthroline disulfonate palladium complex ($PhenS^{*}Pd(OAc)_2$) is employed in water. The authors [2] observed that short-chain alcohols are more soluble, so conversions are faster. Conversely, the reaction rate decreases for longer-chain alcohols because they are less soluble. Using a co-solvent or compounds such as alkanesulfonates or anthraquinone-2-sulfonate provides higher reaction rates. Furthermore, the method can oxidize primary and secondary alyl alcohols, as well as benzyl and aliphatic alcohols with high conversion rates and selectivity.

TEMPO (2,2,6,6-tetramethylpiperidinyl-1-oxy) is a stable nitroxyl radical that can oxidize primary and secondary alcohols smoothly and selectively. For primary alcohols, the oxidation reaction stops at the aldehyde, but the reaction rate is too low possibly because TEMPO coordinates with the metal Pd.

2. Alcohol as Fuel

Interestingly, alcohol can also be employed as fuel to produce energy. This application is restricted to ethanol for toxicity and economic reasons. Ethanol originates from biomass and has reasonable energy content and low toxicity as compared to other alcohols.

The most important example of the successful use of alcohol as fuel is the Brazilian Ethanol Program, which encourages adoption of ethanol as fuel in automobile engines. This Program started almost 50 years ago, and today about 95% of the cars available in the country can run on ethanol or gasoline as fuel because they are equipped with the so-called *flex* engines. Ethanol can be produced from sugar cane, transported around the whole country, and exported to several tropical countries. After an initial criticism on the part of foreign fuel experts, the Brazilian Ethanol Program became widely recognized as an actually less pollutant and renewable energy source as compared to petroleum derivatives, which can be briefly exhausted.

Careful examination of the above-described status of alcohol oxidation to obtain other products or to fuel automobiles leads to the conclusion that any substantial improvement in the processes is somehow limited and hard to achieve. In this scenario, the last important and noticeable enhancement has been to apply green chemistry concepts to accomplish alcohol oxidation to carbonyl compounds. In this scenario, adopting a new approach is desirable, and electrochemistry could be a powerful, clean, environmentally friendly, and efficient tool to oxidize alcohols.

In the next section, we will propose electrooxidation as a tool to overcome the main constraints regarding the development of processes to obtain chemicals and/or energy from alcohol oxidation. The section starts by briefly explaining the ethanol electrochemical characteristics and by further presenting a few examples that demonstrate the alcohol electrooxidation practical features.

3. ALCOHOL ELECTROOXIDATION

From thermodynamic arguments, for any chemical reaction to occur, the chemical potential (μ_c) of the reactant must be increased:

$$\mu_c = \mu^0 + R.T.lna\,,$$

where μ_c is the chemical potential of the species; in other words, the intrinsic chemical potential (μ^0) of a species (structure, number of electrons, number of impaired electrons, charges, dipoles, and other particular characteristics of the species) and the interaction of this species with the medium where it is inserted (*lna*, where *a* is the species activity) determine the tendency of said species to react. Therefore, the own characteristics (structure, number and identity of atoms, electronic density, etc.) and properties of a species under the influence of the medium where it is inserted (interactions with reactants, solvent, and other medium components) determine its tendency to react. From the thermodynamic concepts explained above, we can infer that reactants and experimental conditions influence the alcohol oxidation reaction.

On the other hand but still under thermodynamic arguments, the electrochemical potential of the species (μ_e) determines whether an electrochemical reaction will occur:

$$\mu_e = \mu_c + z.F.\emptyset$$

On the basis of the equation above, any electrochemical reaction will take place when the chemical conditions are fulfilled AND when such reaction relies on additional dependence on the potential $\emptyset$, which is the electrical potential at the local where the electrochemical reaction takes place. The potential $\emptyset$ is further related to the electrode potential E, an experimental parameter that can be externally controlled and acts as driving force for the electrode reaction.

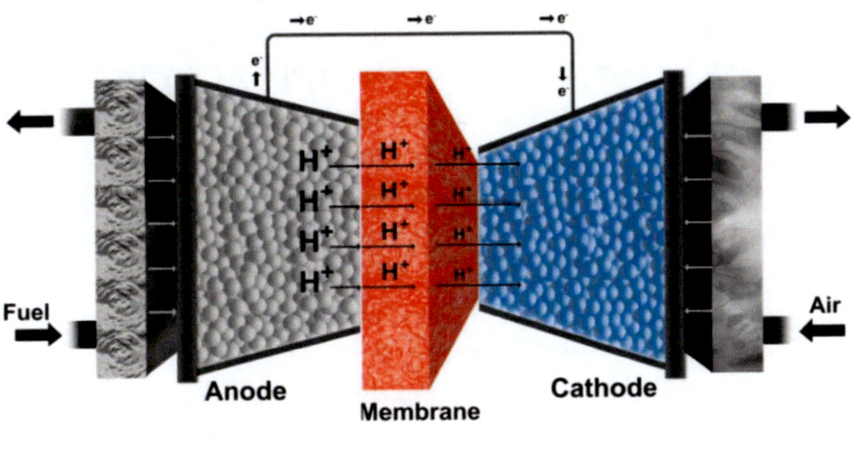

Figure 1. DAFC schematic configuration.

By analyzing the expressions and the consequent thermodynamic implications, we can understand the apparent limitation of chemical alcohol oxidation, which will depend on the reactivity of the alcohol that is in contact with strong oxidizers and on the oxidizing medium. Mild conditions do not usually promote chemical alcohol oxidation. Electrochemistry offers an actual alternative to overcome chemical alcohol oxidation limitations—the electrode potential provides an extra driving force for the reaction to proceed in an environmentally friendly way because the second main reactant is the electron flow. Furthermore, controlled electrode potential leads to selective product formation.

Alcohols have promising use in Direct Alcohol Fuel Cells (DAFCs) [4] (see schematic configuration in Figure 1), which makes them very important subjects for scientific and technological investigations. DAFCs are electrochemical devices that can convert the chemical energy stored in the alcohol molecule chemical bonds to electron flow and, hence, electrical energy. Many alcohols are usually investigated as fuel in DAFCs, but renewable energy sources derived from biomass processes are preferred [4]. Methanol was the first successful experience in terms of alcohol electrooxidation [4]. So far, it has found interesting applications in stationary and mobile systems.

However, methanol is highly toxic, so a few other alcohols have also been evaluated with a view to achieving a renewable and efficient energy source. Ethanol has been exhaustively studied as fuel for DAFCs, but so far not a single electrode material has been able to activate its C-C bond and cleave this alcohol molecule. This restrictive phenomenon usually culminates in acetic acid production during ethanol electrooxidation, which ceases the electrode reaction [5]. Research into improving ethanol oxidation at electrodes is still underway. Ethanol is an attractive fuel for DAFCs because the theoretical number of electrodes transferred from one whole ethanol molecule is higher as compared to methanol, not to mention that ethanol is less toxic and can be easily produced from biomass.

Another attempt to produce energy from alcohol electrooxidation has been to use ethylene glycol (EG). The double alcohol function bound to neighboring C atoms has been hypothesized to activate the EG C-C bond and facilitate its cleavage. Some very interesting results have been obtained with this di-alcohol. We will present some experiences later in this text. Compared to ethanol, a higher number of electrons can be transferred from whole EG oxidation. A further advantage is that EG can be obtained from polyethylene terephthalate (PET) bottles [6], which have become a pollution problem worldwide.

A third initiative concerns glycerol, which has been increasingly produced due to the growing use of biodiesel. A tenth in mass of biodiesel production is glycerol, but the usual industrial processes (cosmetic and animal food production, among others) do not employ glycerol in the same amount it is generated by the biodiesel industry [7, 8]. In Brazil, biodiesel will possibly be mixed in petroleum-derived diesel at a 15% amount in 2025 [9]. In this same country, 3.2 million m^3 of glycerol was produced in 2016 [10]. There is real concern that overproduced glycerol might be inappropriately discarded in nature, causing huge environmental damage. The glycerol molecule can generate 14 electrons per electrooxidized molecule; this alcohol can also originate from renewable biomass.

Table 1 lists the electrochemical characteristics of the alcohols considered here. We will present further details in the next topics.

Two main experimental methods are often employed to assess alcohol electrooxidation performance: cyclic voltammetry, a potentiodynamic technique, and chronoamperometry, a steady-state method [11].

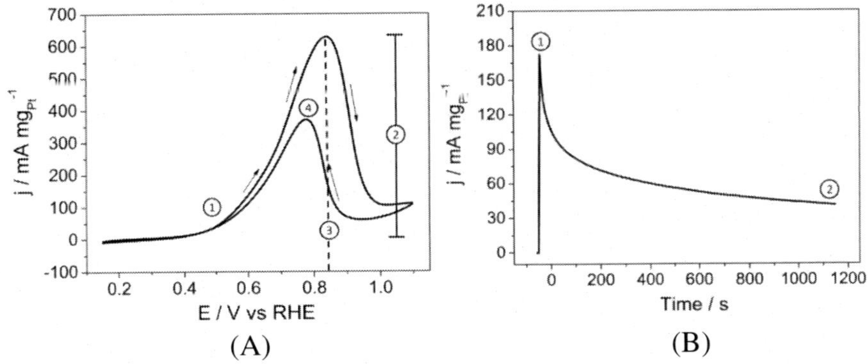

Figure 2. Representative experimental curves for alcohol electrooxidation as evaluated by A) cyclic voltammetry and B) chronoamperometry.

Table 1. Theoretically predicted number of electrons transferred from the total electrooxidation of the alcohol molecule

Molecules	Structures	Number of transferred electrons
Methanol	CH_3OH	6
Ethanol	CH_3CH_2OH	10
Ethylene Glycol	$HOCH_2CH_2OH$	12
Glycerol	$OHCH_2CHOHCH_2OH$	14

Figure 2 illustrates representative experimental curves obtained after application of these two techniques. We will clarify the main points in the curves later in the text.

The curve in Figure 2A shows some points that are considered as the electrode reaction electrochemical performance parameters:

1. Onset potential (OP) – In electrochemistry, the electrode potential (E) corresponds to the energy that is necessary for the electrode process to occur. Therefore, the OP represents the energy that is necessary for the alcohol molecule electrooxidation to start. The lower (less positive) the OP, the less energy is necessary for the oxidation process to begin. This is a thermodynamic parameter that fundamentally depends on alcohol molecule adsorption on the electrode surface.
2. Maximum current density (Ip) – The current results from fuel molecule transformation through the oxidation process. It is a kinetic parameter: the higher the current density, the higher the number of molecules undergoing oxidation. From a practical viewpoint, this value is not reached because mass transport limits the process at this point. In actual systems, lower current values are achieved.
3. Peak potential (Ep) – This is the potential where the maximum electrooxidation process current is reached. Ep is also a thermodynamic parameter because it corresponds to the energy that is necessary for the maximum current to be achieved. Even though the maximum current is not a practical value, the lower the Ep, the

lower the energy necessary to reach significant currents in real systems.

Judging from the aforementioned electrochemical parameters, cyclic voltammetry has become a powerful tool to investigate electrooxidation process efficiency: this technique provides both thermodynamic and kinetic parameters for evaluation. Additionally, an anodic current peak (4) arises in the reverse scan of the potential sweep (negative direction), evidencing that a cleaner process takes place at the electrode surface (removal of products and intermediates) with consequent alcohol molecule re-oxidation thanks to the existence of sufficiently high energy for the process to occur [12]. This is the most characteristic profile of alcohol electrooxidation processes.

To be closer to the actual DAFC, chronoamperometry provides complementary electrochemical information on electrooxidation under stationary conditions. In this technique, an electrode potential (energy) is externally imposed to the system, and the corresponding current (reaction rate) is registered after a while (after the steady-state condition is reached). Figure 2B shows that, after an initial current peak due to oxidation of the molecules present at the electrode surface (1), the current decays and reaches a stationary value (2). This corresponds to the energy that can be extracted at that point. The higher the current extracted at a lower applied E, the better the compound (alcohol) electrooxidation performance. Reaction product identification to support the proposed mechanism and consequent understanding of the whole electrode process often supplement the electrooxidation reaction electrochemical performance evaluation.

In conclusion, the electrooxidation process can be understood differently and provides a few more possibilities than chemical alcohol oxidation. Moreover, new capabilities regarding energy extraction and chemical compound production can be explored. In this perspective, the next section will present a few examples that will aid understanding of the alcohol electrooxidation process.

3.1. Fundamentals of Alcohol Electrocatalysis

Electrochemical processes in general and alcohol electrooxidation in particular strongly depend on the electrode surface where the reaction takes place. The surface plays a crucial role in the reactant molecule adsorption step, which could determine how the reaction proceeds. This adsorption must be strong enough to displace the electronic densities of the bond(s) that will be sequentially broken at the reactant molecule, but it cannot be too strong, or it will make desorption of reaction products and intermediates difficult and will leave a small number of free adsorption sites for the subsequent steps. The first attempts to accomplish alcohol electrooxidation were made with only one metal surface, like Pt, Pd, or Au. Later, electrochemists realized that the combination of metals could elicit a synergistic action and therefore enhance the reaction. Figure 3 exemplifies the phenomena that probably result from the combination of metals as electrode materials for actual adsorption of an alcohol electrooxidation intermediate (CO) on the Pt surface.

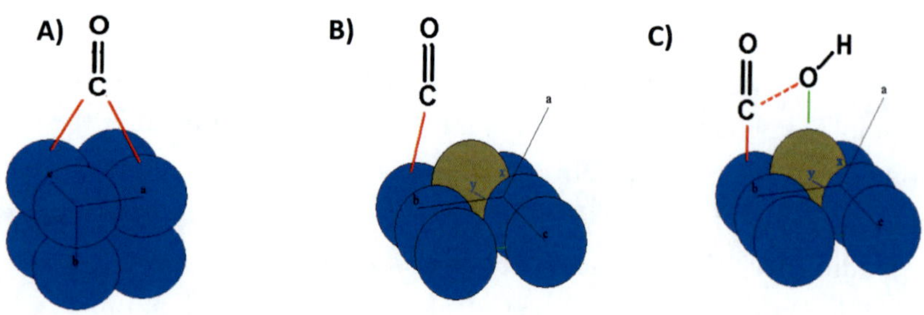

Figure 3. Possible effects caused by insertion of a foreign atom in the Pt matrix on CO adsorption. Blue spheres are Pt atoms; the brown sphere is a foreign metallic atom.

Carbon monoxide is a recognized intermediate in methanol electrooxidation on Pt. This compound passivates the metal surface through irreversible intermediate adsorption, giving rise to a stable bridge configuration on the Pt surface (Figure 3A) [13]. Insertion of the foreign atom could:

- expand the Pt-Pt distance and consequently inhibit CO adsorption in the bridge configuration. Consequently, CO adsorbs in a linear configuration, which is weaker. This is considered the "third body" effect;
- modify the adsorption site (Pt atom) electronic density because the foreign atom is also a chemical species that is different from the matrix and which has distinct electronegativity. As a consequence, the CO-Pt interaction energy varies, resulting in weaker or stronger adsorption. This is considered the real electrocatalytic effect;
- provide an oxygenated species (-OH) in the case of an oxophilic foreign atom. The oxygenated species would then remove CO from the surface via a Langmuir-Hinshelwood mechanism, to produce CO_2. This is called the bi-functional effect.

Methanol oxidation is greatly improved on Pt-Ru surfaces [14] even though there is no consensus about which of the above-mentioned effects or which combination of these effects effectively enhances the process.

At first, this same scientific approach was applied to bulk materials. Today, it is also being applied to nanoparticles.

3.2. Alcohol Electrooxidation Catalyzed by Pt-M Materials

Apart from hydrogen, methanol has also been investigated as fuel in fuel cells because of its promising use as a renewable energy source and high energy/mass ratio. Much research has been devoted to this theme, and actual improvements have been made.

Methanol electrooxidation is known to proceed via the following global reaction:

$$CH_3OH + H_2O \rightarrow CO_2 + 6H^+ + 6e^-,$$

which can occur in two different ways:

$$CH_3OH + Pt \rightarrow Pt\text{-}HCOH + 2H^+ + 2e^- \text{ (Step I)}$$

Step I is common to proceed in both ways. After the HCOH species arises and adsorbs on the Pt surface adsorption site, the reaction may proceed via:

$$Pt\text{-}HCOH + H_2O \rightarrow Pt\text{-}HCO_2H + 2H^+ + 2e^-$$
$$Pt\text{-}HCO_2H \rightarrow Pt + CO_2 + + 2H^+ + 2e^-$$

or

$$Pt\text{-}HCOH \rightarrow Pt\text{-}CO + 2H^+ + 2e^-$$
$$Pt + H_2O \rightarrow Pt\text{-}OH + H^+ + e^-$$
$$Pt\text{-}CO + Pt\text{-}OH \rightarrow 2Pt + CO_2 + H^+ + e^-;$$

water participates in both pathways.

CO (as intermediate) formation in the second pathway is evident. It can be removed by the OH species adsorbed on Pt via a bi-functional effect, which is also considered a Langmuir-Hinshelwood mechanism. CO formation at the Pt surface site is quite problematic because CO can irreversibly adsorb on the Pt surface in a bridged configuration, as mentioned earlier. This constraint has motivated insertion of foreign atoms in the Pt surface, which inhibits the CO poisoning effect.

So far, the PtRu surface has been the most effective in enhancing methanol electrooxidation—the oxophilic Ru atoms present at the Pt surface probably avoid electrode surface poisoning with CO by a functional effect:

$$H_3COH + Pt \rightarrow Pt\text{-}HCOH + 2H^+ + 2e^-$$
$$Pt\text{-}HCOH \rightarrow Pt\text{-}CO + 2H^+ + 2e^-$$
$$Ru + H_2O \rightarrow Ru\text{-}OH + H^+ + e^-$$
$$Pt\text{-}CO + Ru\text{-}OH \rightarrow Pt + Ru + CO_2 + H^+ + e^-$$

Although numerous other surfaces have been developed and tested, until now no noticeable improvement in the electrode reaction has been detected. Furthermore, methanol is highly toxic, which has limited its application and has encouraged the search for alternative alcohol fuels for DAFCs. Anyway, the methanol-fed fuel cell continues to be seen as a very developed and efficient electrochemical device to obtain electrical energy from renewable chemical energy sources.

Ethanol has also been investigated as fuel for fuel cells. Compared to methanol, this alcohol has considerably lower toxicity. Moreover, it is obtained from renewable biomass (most of the time from sugar cane), and it can be easily transported around large countries, from the production plant to the place in need of electrical energy. Still compared to methanol, ethanol can produce $10e^-$ for each complete ethanol molecule electrooxidation via the well-accepted mechanism depicted below:

$$CH_3\text{-}CH_2\text{-}OH \nearrow CH_3CHO \rightarrow CH_3COOH$$
$$CH_3\text{-}CH_2\text{-}OH \searrow C_{1,ads} \rightarrow CO_2$$

The main restriction regarding ethanol electrooxidation is C-C bond activation and molecule splitting, C_1 atom adsorption, and consequent CO_2 production. On Pt surfaces, the reaction follows the CH_3CHO pathway, to give CH_3COOH (acetic acid) as the final main product. Formation of CH_4 as final minor product also occurs at low electrode potentials, which could point to dissociative ethanol molecule adsorption on the surface. After several attempts, $PtSn_x$ surfaces gave the best result. This surface apparently enhances ethanol dissociative adsorption, driving the reaction toward ethanol electrooxidation to CO_2. However, from a practical viewpoint, the electrode materials developed to date do not provide results that are good enough to make ethanol a viable energy source in electrooxidation reaction. Unfortunately, this situation persists and has prevented ethanol from being applied as fuel to obtain electrical energy. Many more improvements in ethanol electrooxidation should be made before this alcohol can be considered a less polluting, easy-to-store and easy-to-transport, inexpensive fuel with high energy potential for use in DAFCs.

Studies involving ethanol have clearly shown the limitation regarding its C-C bond activation and cleavage—the alcohol is completely electrooxidized to CO_2. C_1 carbon undergoes successive dehydrogenations, while C_2 carbon remains unaffected. In this sense, a possible approach to overcome the limitation inherent to ethanol would be to employ di- or tri-alcohols. Therefore, EG and glycerol have emerged as probable alternatives.

EG can originate from acidic hydrolysis and methanolysis during polyethylene terephthalate (PET) recycling reactions [6]. Inappropriate PET bottle disposal has become a serious pollution problem in most countries. These bottles take hundreds of years to decompose under normal environmental conditions and are being produced at an exponential rate. Making this material a source of energy and/or chemicals with high added value would be interesting.

A recent research study evaluated ordered intermetallic PdSb and PtSb nanoparticles [15] as anode material for EG electrooxidation. Cyclic voltammetry and chronoamperometry techniques helped to evaluate the materials for their performance toward the EG electrooxidation reaction; the stable reaction products were identified by Electrochemical Infrared Reflectance Spectroscopy (EIRRS).

Table 2 summarizes the electrochemical parameters obtained for the EG electrooxidation reaction on PtSb/C or PdSb/C ordered intermetallic materials in alkaline medium.

On the basis of the collected data, the PtSb/C nanomaterial has promising performance toward EG electrooxidation in alkaline medium.

Table 2. Electrochemical parameters obtained for EG electrooxidation on different electrode materials, in alkaline medium, at room temperature. The data were obtained by cyclic voltammetry. Scan rate = 20 mV.s^{-1}. RHE is reversible hydrogen electrode.

	Onset potential (V vs RHE)	Peak current density (mA $mg_{Pt/Pd}^{-1}$)	Oxidation peak potential (V vs RHE)
Pt/C	0.38	38.8	0.83
PtSb/C	0.40	22,1	0.74
PdSb/C	0.49	6.8	0.83

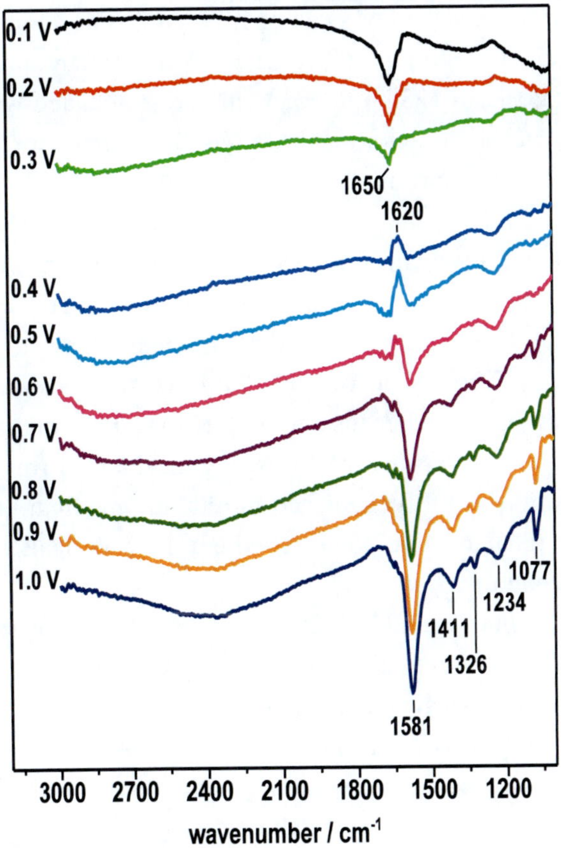

Figure 4. *In situ* EIRRS spectra, collected in the reflectance mode, recorded for PtSb/C surface polarized at different electrode potentials toward EG electrooxidation in alkaline medium at room temperature.

The products generated during the electrooxidation reaction were identified by infrared spectroscopy (Electrochemical Infrared Reflectance Spectroscopy – EIRRS); Figure 4 contains the spectra of each selected electrode potential.

The spectra in Figure 4 display IR absorptions at 1650 and 1642 cm^{-1}, which correspond to adsorbed water angular bending [16]. The peaks at 1587 and 1581 cm^{-1} refer to glyoxal, formate ion, glycolate, and glyoxalate (COO^-) or (C-O) asymmetric stretching [17] or even only to glycolate [18]. The peak at 1441 cm^{-1} is assigned to carbonate ion [19] and also to glycolate asymmetric stretching [17]. The peak at 1326 cm^{-1} is attributed to glycolate ion (COO^-) symmetric stretching [17]. However, Demarconnay ascribed it to oxalate ions. Peaks at 1235 and 1234 cm^{-1} are associated with carboxylic acid (C-O) symmetric stretching [18]. Glycolic acid is most probably the main product of EG electrooxidation because it contributes to the peak at 1234 cm^{-1} the most. The peaks assigned to glycolate ion at 1581, 1411, and 1326 cm^{-1} corroborate such conclusion. On the other hand, oxalate ion formation cannot be discarded because the peak at 1235 cm^{-1} corresponds to oxalic acid (C-OH) stretching [20, 21]. Additionally, the peak at 1411 cm^{-1} confirms carbonate formation; however, this is not considered the main electrode reaction pathway because peaks due to CO_2 are absent.

Considering all the results, the following mechanism was proposed for EG electrooxidation at PtSb/C surface in alkaline medium. The electrode potentials E are related to the potential where the chemicals were detected by the EIRRS technique.

Even though complete EG electrooxidation was not achieved; that is, the total number of theoretically intended electrons was not obtained, the electrooxidation process generated a series of chemical compounds with great added value (Table 4). Hence, EG electrooxidation is a promising process where the alcohol compound originally obtained from the recycled PET bottles can be used as fuel to produce energy and chemicals of interest in the chemical industry.

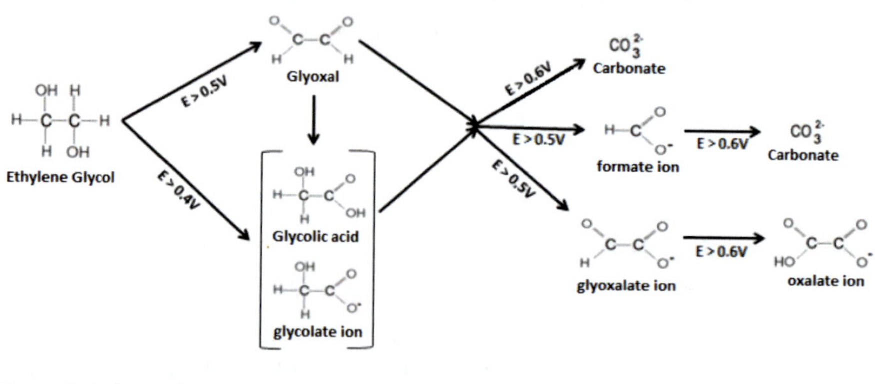

Figure 5. Schematic mechanism proposal for ethylene glycol electrooxidation at PtSb/C surface in alkaline medium.

As pointed out earlier, glycerol has been produced in much greater amount than it is used in the cosmetic, animal food, and pharmaceutical industries. This situation increases the risk of inappropriate disposal in the environment of this co-product generated during biodiesel production. Worldwide, countless researchers have joined efforts to find innovative ways to use this compound, including electrochemists. Complete electrooxidation of one glycerol molecule to CO_2 and H_2O potentially produces 14 electrons. Therefore, this tri-alcohol has been investigated as fuel for DAFCs.

The same electrode materials tested for EG electrooxidation were also evaluated for glycerol electrooxidation in alkaline medium. Table 3 displays the electrochemical parameters collected from the assays.

Steady-state measurements corroborated the set of promising results. PtSb/C and commercially available Pt/C reached maximum current densities of 3.7 mA/mg of Pt and 2.4 mA/mg of Pt, respectively.

The PtSb/C material performance toward glycerol electrooxidation was clearly outstanding and surpassed Pt/C performance under the same experimental conditions, encouraging the use of the tri-alcohol as energy source. The main electrooxidation reaction products were identified by EIRRS (Figure 6).

Table 3. Electrochemical parameters obtained for glycerol electrooxidation at different electrode materials, in alkaline medium, at room temperature. Data obtained from cyclic voltammetry technique. Scan rate = 20 mV.s^{-1}

	Onset potential (V vs RHE)	Peak current density (mA mg $_{Pt,/Pd}$ $^{-1}$)	Oxidation peak potential (V vs RHE)
Pt/C	0.37	18.7	0.80
PtSb/C	0.40	25.6	0.75
PdSb/C	0.45	10.4	0.85

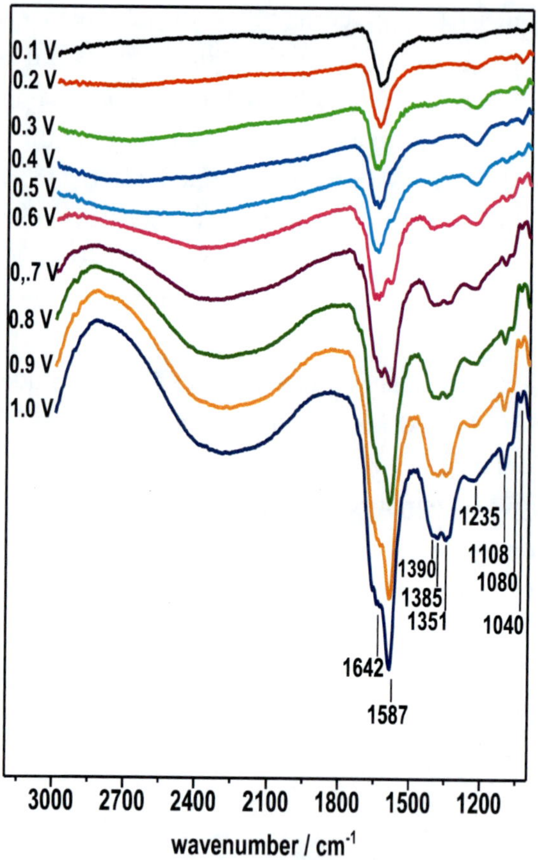

Figure 6. *In situ* EIRRS spectra collected in the reflectance mode of the PtSb/C surface polarized at different electrode potentials during glycerol electrooxidation in alkaline medium at room temperature.

The peak at 1235 cm^{-1} concerns carboxylic acid and/or glycolate ion formation from 0.2V. Hence, the glycerol C-C bond cleaves at remarkably low electrode potentials, indicating that low energy is necessary to initiate C-C bond activation in this molecule. Still at 0.2V, the peak at 1040 cm^{-1} refers to glycoaldehyde and glycerol molecules. At potentials more positive than 0.6V, this peak assumes a positive profile, suggesting that glycerol depletion from the thin film on the electrode surface overcomes glycolaldehyde production because of the high energy applied to the electrode. For potentials higher than 0.4V, the peak at 1108 cm^{-1}

corresponds to glycolaldehyde and glyoxalate ion. From 0.5V, there is a peak at 1580 cm^{-1}, assigned to glyoxal, formate, glycolate, and glyoxalate ions, as well as a peak at 1400 cm^{-1}, attributed to glycolate, glycerate, and carbonate ions. The peaks due to hydroxy-piruvate and formate ions (1350 cm^{-1}) and glycolate ion (1040 cm^{-1}) emerge at potential 0.6V and higher.

On the basis of the results described above, the mechanism was proposed for glycerol electrooxidation at PtSb/C nanoparticles in alkaline medium:

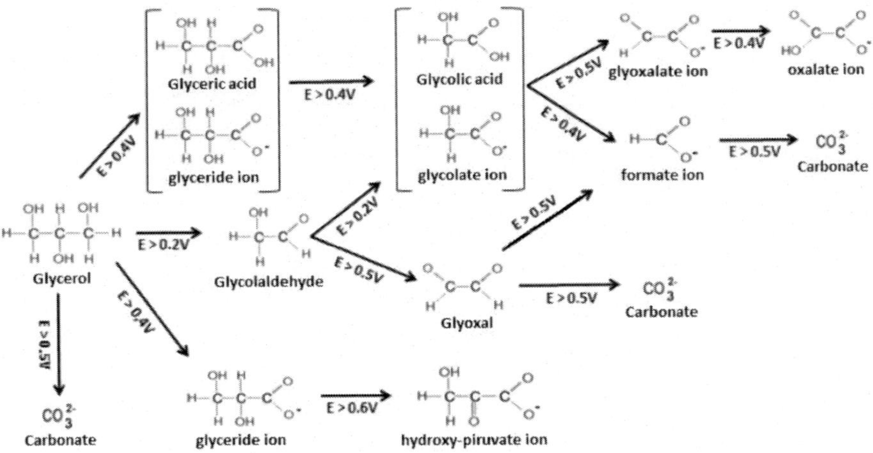

Figure 7. Schematic mechanism proposal for glycerol electrooxidation at PtSb/C surface in alkaline medium.

From the last examples, EG and Glycerol must be seriously considered as promising fuels for electrical energy production through DAFCs. Even though the electrode reaction electrical efficiency is still not too high, both organic compounds come from waste that can potentially pollute the environment, so their transformation in anything useful is a big deal. This issue, *per se*, justifies interest in alcohol electrooxidation, and research in this field will certainly enhance electrode reaction efficiency and yield better materials for electrocalysis. However, a deeper analysis of the mechanisms proposed for the reactions in Figures 5 and 7 uncovers a series of chemical compounds that can be of great benefit and interest to different chemical industries. Electrode potential control along the electrochemical process and an appropriate method of separation can provide electrical energy and valuable chemicals simultaneously.

Table 4 shows the most important chemicals generated during EG and glycerol electrooxidation and the corresponding prices in the international market [22-24]. Some compounds may make the electrochemical process viable from an economic viewpoint if they are properly enhanced.

Table 4. Commercial prices of some products obtained from EG and glycerol electrooxidation in alkaline medium

Alcohol	Intermediate/product	Price (USD)
Ethylene glycol	Glyoxal	~ 1500/kg
Ethylene glycol	Glycolic acid	250/kg
Ethylene glycol	Glyoxalate	170/g
Glycerol	Glyceric acid	315/g
Glycerol	Hydroxypyruvic acid	73/g
Glycerol	Glycolaldehyde	90/g

Conclusion

Alcohols present interesting properties for energy production and synthesis of chemical products. Chemical alcohol oxidation can yield both energy and valuable products. In this chapter, we showed that the alcohol oxidation electrochemical route, called electrooxidation, can be a powerful tool to explore all the potentialities of the alcohol reaction. Like any heterogeneous reaction, alcohol electrooxidation heavily depends on the development of surfaces where the reaction can occur. This is a broad and still incipient research field that needs to undergo extensive development, especially after the advent of nanotechnology. Alcohol electrooxidation will undoubtedly contribute to energy generation, chemical industry, and the environment, among other fields.

REFERENCES

[1] Bruice, P. Y. (2006). *Química Orgânica.*, [*Organic chemistry*]. 263-266, São Paulo: Pearson Prentice Hall.

[2] Brink, G-J ten., Arends, I. W. C. E. & Sheldon, R. A. (2000). "Green, Catalytic Oxidation of Alcohols in Water." *Science*, *287*, 1636-1639. Accessed July 10, 2018. doi: 10.1126/science.287.5458.1636.

[3] Ji, H. B., Ebitani, K., Mizugaki, T. & Kaneda, K. (2002). "Environmentally friendly alcohol oxidation using heterogeneous catalyst in the presence of air at room temperature." *Catalysis Communications*, *3*, 511–517. Accessed July 10, 2018. doi:10.1016/S1566-7367(02)00194-2.

[4] Vielstich, W., Lamm, A. & Gasteiger, H. A. (2003). "Part 6: Oxidation of small organic molecules" in *Handbook of Fuel Cells: Fundamentals Technology and Applications*, 601-678. England: Wiley.

[5] Vigier, F., Countaceau, C., Hahn, F., Belgsir, E. M. & Lamy, C. (2004). "On the mechanism of ethanol electro-oxidation on Pt and PtSn catalysts: electrochemical and in situ IR reflectance spectroscopy studies." *Journal of Electroanalytical Chemistry*, *563*, 81-89. Accessed August 20, 2018. doi.org/10.1016/j.jelechem.2003.08.019.

[6] Brandrup, J., Bittner, M., Michaeli, W., Menges, G. & Willenberg, B. (1996). *Recycling and recovery of plastics*, Munich: Hanser Gardners.

[7] Beatriz, A., Araújo, Y. J. K. & LIMA, D. P. (2011). "Glicerol: Um breve histórico e aplicação em sínteses estereosseletivas." [Glycerol: A brief history and application in stereoselective syntheses]. *Química Nova*, *34*, 306-319. Accessed August 29, 2018. http://dx.doi.org/10.1590/S0100-40422011000200025.

[8] Yang, F., Hanna, M. A. & SUN, R. (2012). "Value-Added Uses for Crude Glycerol- a Byproduct of Biodiesel Production." *Biotechnology for Biofuels*, *5*,1 10. Accessed August 29, 2018. doi.org/10.1186/1754-6834-5-13.

[9] Trigueirinho, F., Minelli, J. C. & Tokarski, D. (2016). "Biodiesel: oportunidades e desafios no longo prazo". *Brasília*, October 06. Accessed August 23, 2018. http://www.abiove.org.br/site/

_FILES/Portugues/07102016-131231-07_10_2016_n-_cenario_para_o_biodiesel_em_2030(2).pdf.

[10] Portal, Brasil. (2017). "*Produção de biodiesel atinge recorde de 351 mil metros cúbicos*"., [*Biodiesel production reaches a record 351 thousand cubic meters*]. January 10, Accessed August 23, 2018. http://www.brasil.gov.br/editoria/infraestrutura/2017/01/producao-de-biodiesel-atinge-recorde-de-351-mil-metros-cubicos.

[11] Bard, A. J. & Faulkner, L. R. (2001). *Electrochemical Methods: Fundamentals and Applications*. New York: Wiley.

[12] Iwasita, T. (2002). "Fuel Cells: Spectroscopic Studies in the Electrocatalysis of Alcohol Oxidation." *Journal of the Brazilian Chemical Society*, *13*, 401-409. Accessed August 29, 2018. http://dx.doi.org/10.1590/S0103-50532002000400002.

[13] Ângelo, A. C. D. (2016). "The Role of Electrochemistry in a Cleaner World ". In *Advances in Cleaner Production*, edited by Giannetti, B. F., Almeida, C. M. V. B., Agostinho, F., 319-336. New York: Nova Science Publishers.

[14] Iwasita, T., Hoster, H., John-Anacker, A., Lin, W. F. & Vielstich, W. (2000). "Methanol Oxidation on PtRu Electrodes. Influence of Surface Structure and Pt−Ru Atom Distribution". *Langmuir*, *16*, 522-529. Accessed August 29, 2018. DOI: 10.1021/la990594n.

[15] Scachetti, T. P. & Ângelo, A. C. D. (2015). "Ordered Intermetallic Nanostructured PtSb/C for Production of Energy and Chemicals". *Electrocatalysis*, *6*, 472−480. Accessed August 28, 2018. doi.org/10.1007/s12678-015-0265-y.

[16] Beden, B., Mobin, M. C., Hahn, F. & Lamy, C. (1987). "*In Situ*" Analysis by Infrared Reflectance Sprectroscopy of the Adsorbed Species Resulting from the Electrosorption of Ethanol on Platinum in Acid Medium". *Journal of Electroanalytical Chemistry*, *229*, 353–366. Accessed August 28, 2018. doi.org/10.1016/0022-0728(87)85152-5.

[17] Chang, S. C., Ho, Y. & Weaver, J.M. (1991). "Applications of Real-Time FTIR spectroscopy to the elucidation of complex electroorganic pathways: Electrooxidation of ethylene glycol on gold, platinum, and

nickel in alkaline solution". *Journal American Chemical Society*, *113*, 9506-9513. Accessed August 28, 2018. DOI: 10.1021/ja00025a014.

[18] Demarconnay, L., Brimaud, S., Coutanceau, C. & Léger, J. M. (2007). "Ethylene glycol electrooxidation in alkaline medium at multi-metallic Pt based catalysts". *Journal of Electroanalytical Chemistry*, *601*, 169-180. Accessed August 28, 2018. doi.org/10.1016/j.jelechem. 2006.11.006.

[19] Christensen, P. A. & Hamnet, A. (1989). "The oxidation of ethylene glycol at platinum electrode in acid and base. An in situ FTIR study". *Journal Electroanalytical Chemistry*, *260*, 347-359. Accessed August 28, 2018. doi.org/10.1016/0022-0728(89)87149-9.

[20] Leung, L. W. H. & Weaver, M. J. (1990). "Influence of Adsorbed Carbon Monoxide on the Electrocatalytic Oxidation of Simple Organic Molecules at Platinum and Palladium Electrodes in Acidic Solution: A Survey Using Real-Time FTIR Spectroscopy". *Langmuir*, *6*, 323-333. Accessed August 29, 2018. DOI: 10.1021/la00092a006.

[21] Cabaniss, S. E., Leenheer, J. A. & McVEY, I. F. (1998). "Aqueous infrared carboxylate absorbances: aliphatic di-acids". *Spectrochimica Acta Part A*, *54*, 449-458. Accessed August 29, 2018. doi.org/10.1016/S1386-1425(97)00258-8.

[22] Alibaba – Global trade starts here. (2018). "*Products*". Accessed August 20. https://www.alibaba.com/.

[23] CarlRoth. (2018). "*Search for description*". Accessed August 20. https://www.carlroth.com/en/en.

[24] Molbase. (2018). "*Search*". Accessed August 20. http://www.molbase.com/.

Biographical Sketch

Antonio Carlos Dias Ângelo, PhD

Affiliation: UNESP - Sao Paulo State University.

Education: Bachelor in Chemistry University of Sao Paulo – USP; Masters and PhD in University of Sao Paulo – USP; Associate Professor in UNESP – Sao Paulo State University.

Business Address: Av. Eng. Luiz Edmundo Carrijo Coube, 1401. 17033-360. Bauru, SP, Brazil.

Research and Professional Experience: Associate Professor at UNESP - São Paulo State University for 30 years. Pos-doctor fellow at Sherbrooke University, CANADA. Visiting Scientist at Cornell University.

Publications from the Last 3 Years:

[1] Bortoloti, Francielle., Ishiki, Nicolas Andrade., Della-Costa, Maria-Laura., Rocha, Kleper Oliveira. & Angelo, Antonio C.D. (2018). The Influence of Pt-Sn System Nanostructure on the Electronic Condition of Pt Adsorption Surface Site. *Journal of Physical Chemistry C.*

[2] Bortoloti, F. & Angelo, A. C. D. (2017). Ordered PtSn/C Electrocatalyst: A High Performance Material for the Borohydride Electrooxidation Reaction. *Catalysts.*, v. *7*, p. 198.

[3] Bortoloti, F., Garcia, A. C. & Angelo, A. C. D. (2015). Electronic effect in intermetallic electrocatalysts with low susceptibility to CO poisoning during hydrogen oxidation. *International Journal of Hydrogen Energy.*, v. *40*, p. 10816 - 10824.

[4] Scachetti, T. P. & Angelo, A. C. D. (2015).Ordered Intermetallic Nanostructured PtSb/C for Production of Energy and Chemicals. *Electrocatalysis* (Print), v. *6*, p. 472 – 480.

[5] Perini, Nickson., Batsta, B., Angelo, A. C. D., Epstein, I. & Varela, Hamilton. (2014).Long-Lasting Oscillations in the Electro-Oxidation of Formic Acid on PtSn Intermetallic Surfaces. *ChemPhysChem.*, v.*15*, p.1753 - 1760.

[6] Pinto, L.M.C., Juárez, M.F., Ângelo, A.C.D. & Schmickler, W. (2014). Some properties of intermetallic compounds of Sn with noble metals relevant for hydrogen electrocatalysis. *Electrochimica Acta.*, v.*116*, p.39 - 43.

[7] Kushikawa, Ricardo T., Silva, Marcelo R., Angelo, Antonio C.D. & Teixeira, Marcos F.S. (2016). Construction of an electrochemical sensing platform based on platinum nanoparticles supported on carbon for tetracycline determination. *Sensors and Actuators. B, Chemical.*, v.*228*, p.207 - 213.

[8] Angelo, A.C.D. (2016). *Advances in Cleaner Production vol. 2 - The Role of Electrochemistry in Cleaner World*. Hauppauge NY: Nova Science Publishers, v.*1*, p.18.

INDEX

#

α

δ

A

B

C

D

E

F

G

H

I

K

L

M

N

O

P

Q

R

S

T

The Physicochemical Aspects of Nitric Oxide in Chemistry and Biology: Fundamentals and Recent Developments

Author: Gertz I. Likhtenshtein, Ben-Gurion University of the Negev, Beer-Sheva Israel and Institute of Problems of Chemical Physics, Russian Academy of Science, Chernogolovka, Moscow region, Russian Federation

Series: Chemistry Research and Applications

Book Description: This book represents a view on the area from physical chemists with long and broad expertise in chemical biophysics, mechanisms of enzyme catalysis and physical methods for application in biology, in particular.

Hardcover ISBN: 978-1-53613-958-7
Retail Price: $230

Chemical Elements (Fluorine, Rhodium and Rubidium): Properties, Synthesis and Applications

Editor: Adrian Huff

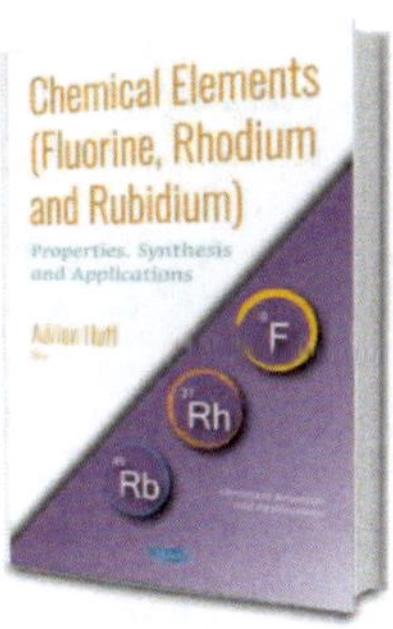

Series: Chemistry Research and Applications

Book Description: In this collection, controlled fluorination was successfully applied for the fluorination of different starting materials: SWCNTs, MWCNTs, carbon nanodiscs, carbon blacks, polymers, carbides, and Li4Ti5O12 (electrochemical materials). The choice of fluorinating agent is crucial as it defines the temperature range as well as the nature of the decomposition products.

Softcover ISBN: 978-1-53614-017-0
Retail Price: $95

Acetic Acids: Advances in Research and Applications

Editors: Mariusz Szymczak, Department of Food Science and Technology, West Pomeranian University of Technology, Szczecin, Poland; Osman Topuz, Akdeniz University, Fisheries Faculty, Seafood Processing Department, Antalya, Turkey

Series: Chemistry Research and Applications

Book Description: This book covers the problems related to acetic acid in the food industry. People have been using acetic acid to produce food for thousands of years and continue to find new uses for it. This book is – in principle – intended for students and staff of agricultural academies, but may also be of interest to food producers who will find in it a number of tips for food technology related to the use of acetic acid.

Hardcover ISBN: 978-1-53613-550-3
Retail Price: $160

Terpenes: Biosynthesis, Applications and Research

Editors: Albin Bjarke

Series: Chemistry Research and Applications

Book Description: In this collection, terpenes in oregano, mainly represented by the essential oils, are extensively documented for their antimicrobial, antifungal, antiviral, antioxidant, anti-inflammatory and anticancer properties, with the goal of describing the biological properties of terpenes from oregano and their mechanism of action.

Softcover ISBN: 978-1-53613-622-7
Retail Price: $95